I0482785

NUREG/CR–6334

New Sensor for Measurement of Low Air Flow Velocity

Phase I Final Report

Manuscript Completed: June 1995
Date Published: August 1995

Prepared by
H. M. Hashemian, M. Hashemian, E. T. Riggsbee

Analysis and Measurement Services Corporation
AMS 9111 Cross Park Drive
Knoxville, TN 37923

Prepared for
Division of Regulatory Applications
Office of Nuclear Regulatory Research
U.S. Nuclear Regulatory Commission
Washington, DC 20555–0001
NRC Job Code W6378

ISBN-13: 978-1499576955; ISBN-10: 1499576951

This manuscript has been authored by a contractor of the U.S. Government under Grant No. W6378. Accordingly, the U.S. Government has a nonexclusive, royalty-free license to publish or reproduce the published form of this contribution, or allow others to do so, for U.S. Government purposes.

ABSTRACT

Personnel radiation protection in nuclear facilities requires information about the speed and direction of air flow in the work area. This information is needed to determine where to locate air samplers to collect airborne radioactive material that radiation workers may inhale.

Conventional flow sensors are not sensitive enough for the low air flow rates that are usually involved in radiation protection applications. Furthermore, a practical means is not currently available to determine the direction of air flow. Smoke candles and isostatic bubbles that are presently used for detecting air flow patterns have several shortcomings and new methods are needed to provide more reliable results.

The project described here is the Phase I feasibility study of a two-phase program to integrate existing technologies to provide a system for determining air flow velocity and direction in radiation work areas. Basically, a low air flow sensor referred to as a thermocouple flow sensor has been developed. The sensor uses a thermocouple as its sensing element. The response time of the thermocouple is measured using an existing in-situ method called the Loop Current Step Response (LCSR) test. The response time results are then converted to a flow signal using a response time-versus-flow correlation.

The Phase I effort has shown that a strong correlation exists between the response time of small diameter thermocouples and the ambient flow rate. As such, it has been demonstrated that thermocouple flow sensors can be used successfully to measure low air flow rates that can not be measured with conventional flow sensors.

While the thermocouple flow sensor developed in this project was very successful in determining air flow velocity, determining air flow direction was beyond the scope of the Phase I project. Nevertheless, work was performed during Phase I to determine how the new flow sensor can be used to determine the direction, as well as the velocity, of ambient air movements. Basically, it is necessary to use either multiple flow sensors or move a single sensor in the monitoring area and make flow measurements at various locations sweeping the area from top to bottom and from left to right. The results can then be used with empirical or physical models, or in terms of directional vectors to estimate air flow patterns. The measurements can be made continuously or periodically to update the flow patterns as they change when people and objects are moved in the monitoring area. The potential for using multiple thermocouple flow sensors for determining air flow patterns will be examined in Phase II.

TABLE OF CONTENTS

LIST OF FIGURES

LIST OF TABLES

FOREWORD

NUREG/CR-6334 is not a substitute for NRC position papers or regulations, and compliance is not required. The results, approaches, and methods described in this NUREG are provided for information only. Publication of this report does not necessarily constitute NRC approval or agreement with the information contained herein.

John E. Glenn, Chief
Radiation Protection and
Health Effects Branch
Division of Regulatory Applications
Office of Nuclear Regulatory Research

ACKNOWLEDGEMENTS

The cooperation of a number of individuals and organizations is gratefully acknowledged.

The Engineering Technology Division of the Oak Ridge National Laboratory (ORNL) provided their Laser Doppler Velocimeter (LDV) for air flow measurements in the project. This was arranged through a technology transfer agreement with Martin Marietta Energy Systems, Inc. which operates ORNL for the Department of Energy. Mr. David K. Felde and Mr. George Farguharson from ORNL assisted AMS in the use of the LDV.

Mr. Steve Norris and Mr. Michael Lauer of Scientific Ecology Group (SEG), Inc. of Oak Ridge, Tennessee helped arrange for a site visit of AMS to their facilities to observe their air sampling practices.

Also, the cooperation of Mr. Wayne Knox of Advanced System Technology in Atlanta, Georgia, Mr. J. Mishima of SAIC in Richland, Washington, and Mr. Andrew M. Maxin of Nuclear Fuel Services, Inc. in Erwin, Tennessee is acknowledged.

Mr. Frank Hahne and Mr. Henry Bailey of NFS Radiation Protection Systems in Norcross, Georgia have been helpful during the Phase I project and have agreed to cooperate with AMS during the Phase II effort.

1. INTRODUCTION

This report presents the results of a Phase I research project conducted over a six month period for the U.S. Nuclear Regulatory Commission (NRC). The purpose of the project was to determine the feasibility of a new sensor for the measurement of ambient air movements to aid in proper placement of air samplers for personnel radiation protection applications in nuclear fuel facilities.

Air sampling is performed in nuclear facilities to control and minimize the radiation exposure to workers and to comply with NRC requirements outlined in 10CFR Part 20, "Standard For Protection Against Radiation."[1]

Air sampling in most nuclear facilities is currently performed using fixed location air samplers, continuous air monitors, portable samplers, and lapel samplers. Experience has shown that the airborne concentrations measured by these devices can vary widely within an area due to the random distribution of airborne materials[2]. Therefore, it is crucial for air samplers to be placed in proper locations to collect representative samples of airborne material inhaled by radiation workers. This requires an understanding of both air flow direction and velocity in the work area.

Air flow direction is usually determined using smoke candles to visually observe air flow direction. This technique is effective only in very low flow regions. The smoke diffuses too rapidly to allow tracking by observation unless the velocity is very low. Smoke is also a respiratory irritant; therefore, workers and personnel have to wear full-face respirators when the smoke is present. In addition, all sensitive equipment in the area must be covered for protection from the smoke residue.

The new flow sensor described in this report uses a thermocouple as its sensing element. As such, it is called a thermocouple flow sensor. Exposed junction or small diameter sheathed thermocouples of any types can be used to construct the new flow sensor. In the Phase I project reported here, Type K thermocouples were used to construct laboratory prototype sensors, but other thermocouple types should be equally suitable. The response time of the thermocouple which is very sensitive to flow, is measured and converted to a flow signal using a response time-versus-flow correlation. The response time is measured by a new method called the Loop Current Step Response (LCSR) test. This method was originally developed for in-situ measurement of response times of Resistance Temperature Detectors (RTDs) in nuclear power plants. It was later adapted for measurement of response times of thermocouples in aerospace applications. The advantages of the LCSR method is that it can be used remotely to measure the response time of a thermocouple under the conditions that the thermocouple is used.[3,4]

The idea for the new thermocouple flow sensor proposed here was conceived in the early 1980s. However, at that time, the LCSR method for in-situ measurement of thermocouple response time was not as advanced as it is today.

The LCSR test is based on heating the thermocouple with an electric current applied remotely at the end of thermocouple extension wires. The current is applied for a few seconds to heat the thermocouple and raise its temperature several degrees above the ambient temperature. The current is then cut off and the transient output of the thermocouple is recorded as it returns to the ambient temperature. This transient can be analyzed to provide the response time of the thermocouple under the conditions tested.

The analysis of the LCSR transient involves a sophisticated mathematical transformation and a computer fitting technique to convert the internal heating response of the thermocouple to the transient response that would be obtained if the thermocouple was exposed to a sudden change in the ambient temperature. The transient response to a sudden change in the ambient temperature is the data which is conventionally used to identify the response time of the thermocouple in terms of a time constant. The time constant is defined as the time required for a sensor output to reach 63.2 percent of its final steady-state value following a step change in the ambient temperature. The details of the LCSR method are included in this report.

Following is a list of tasks that have been completed in Phase I to develop a thermocouple flow sensor (Table 1.1).

1. LCSR test equipment remaining from previous R&D work was modified, reassembled, and set up to perform in-situ response time measurements on thermocouples as necessary for the Phase I work. Some software modifications were also performed to adapt the LCSR software to thermocouples for the purpose of flow measurements.

2. An air flow loop was constructed using a stepper motor to operate a fan to provide air flow rates from near stagnant up to 2 meters per second. The flow loop was then transported to Oak Ridge National Laboratory (ORNL) where a Laser Doppler Velocimeter was used to calibrate the loop. The calibration involved measuring air flow rates in the loop as a function of stepper motor speed setting.

3. The laboratory flow loop was used to generate experimental response time-versus-flow data for a number of type K thermocouples. Using experimental data, an empirical correlation was developed to give air flow rate as a function of thermocouple response time for a given thermocouple size.

4. An informal survey involving site visits, interviews, telephone contacts, and personal contacts was made with a number of individuals and organizations in the field of radiation protection.

5. In an effort to address flow direction in addition to flow velocity, work was started in Phase I to determine ways to use flow velocity information to establish flow patterns. It was determined that the key to mapping of flow in a room is to use multiple thermocouples in various locations and measure air flow rates. The flow rate data can then be analyzed using theoretical modeling to establish the room's flow profile. The feasibility of this method for air flow pattern determination will be examined in Phase II.

TABLE 1.1

Summary Of Phase I Accomplishments

---

1. Adaptation of existing LCSR equipment and software for use with thermocouples to measure air flow velocity

2. Design and construction of a laboratory air flow loop and calibration of the loop using a Laser Doppler Velocimeter at Oak Ridge National Laboratory

3. Development of response time-versus flow data for several thermocouples and generation of empirical correlations

4. Survey of nuclear fuel industry including site visits, interviews with experts in the field, and review of related literature

5. Investigation of theoretical techniques that can be combined with thermocouple flow measurements to estimate air flow patterns

6. Contact with experts in the field of Computational Fluid Dynamics (CFD) and neural networks for indoor air flow mapping and determination of air flow patterns

---

2. TECHNICAL BACKGROUND

Numerous sensors are commercially available for the measurement of air velocity in a variety of industrial and scientific applications. However, these sensors do not have adequate sensitivity, accuracy, and reliability for measurement of very low flow rates (less than 0.5 meters per second).

A smooth correlation naturally exists between the response time of most thermocouples and the flow rate of the environment in which the thermocouple is installed. Figure 2.1 illustrates the typical response time of an exposed-junction thermocouple as a function of flow rate in ambient air. It is apparent that the response time is extremely sensitive to flow rate at low flows, and that the sensitivity decreases as the flow rate increases. The high sensitivity of a thermocouple response time at low air flows is the reason why thermocouple flow sensors are successful in measuring very low flow rates.

Figure 2.2 provides a comparison between the sensitivity of a thermocouple flow sensor and a typical conventional flow meter. It is apparent that the thermocouple flow sensor is very sensitive at low flows where conventional flow sensors have poor sensitivity. Conversely, conventional flow sensors have better sensitivity at high flow rates than thermocouple flow sensors. That is, thermocouple flow sensors can help overcome an important shortcoming of present flow sensing devices.

The following procedure outlines how a thermocouple may be converted into a low flow rate sensor.

1. Select a fast-response thermocouple (e.g., a 1/64" exposed-junction Type K thermocouple). The thermocouple type being K, J, T, or others is not as important in this application as the thermocouple size, especially the size of the junction. This is because the response time of a thermocouple is essentially independent of type, but is very sensitive to the thermocouple size.

2. Measure the response time of the thermocouple at three or more widely spaced, but low flow rates in air. The response time measurement can be performed using the LCSR method or the conventional plunge test method as described later in this report.

3. Fit the data measured in step 2 above to a suitable correlation or a logarithmic series equation which best defines the relationship between the response time of a thermocouple and the flow rate of the surrounding media. The general form of such correlations is given later in this report along with the derivation of the correlation. The logarithmic series equation may be obtained by an empirical approach.

4. Install the thermocouple in the media whose flow is to be measured and connect the thermocouple's extension leads to the LCSR test equipment.

5. Measure the response time of the thermocouple whenever the flow information is needed, and use the correlation from step 3 to convert the response time to flow rate information.

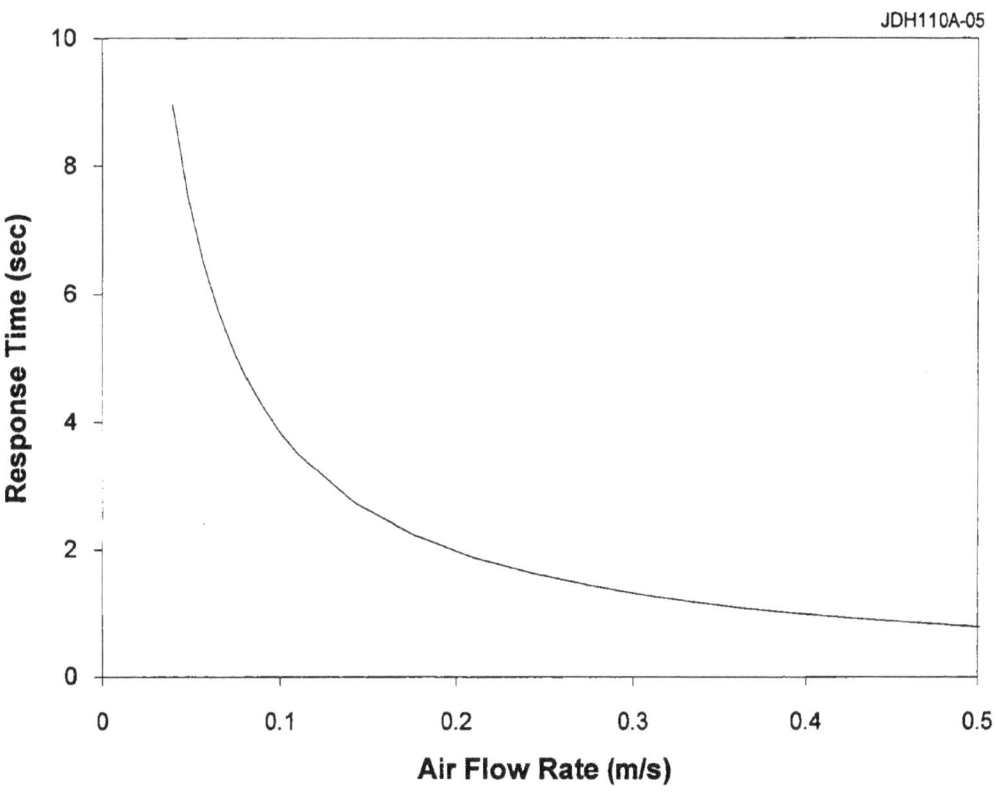

Figure 2.1 Typical Correlation Between Response Time of a
Type K Thermocouple and Air Flow Rate

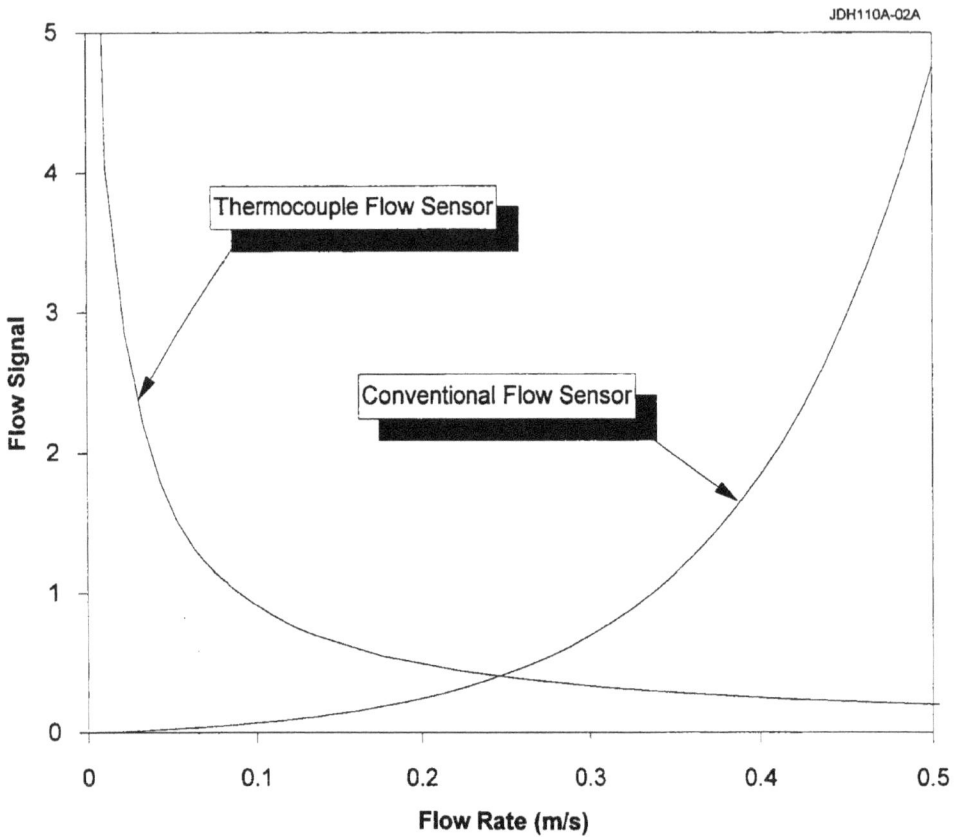

Figure 2.2 Comparison of Sensitivity of a Typical Thermocouple
Flow Sensor and a Conventional Flow Sensor

The feasibility of the above approach was established in this Phase I project. In Phase II, commercial prototype equipment will be designed and constructed for automatic measurement of low air flow rates. Furthermore, determining air flow direction will be addressed in Phase II.

In Phase I, it was determined that multiple thermocouple flow sensors can be used in a radiation work area to map the air flow rates in the area. This information can then be used with empirical or physical models to predict air flow patterns. Empirical modeling using neural networks will be attempted for this application in Phase II. In addition, physical models based on computational fluid dynamics (CFD) and other mathematical techniques were identified in Phase I that have the potential to be used with velocity and temperature data from multiple thermocouple flow sensors to predict flow patterns. These models have been successfully used for air flow mapping for such applications as determining where to locate smoke detectors in residential and commercial buildings. The CFD type models and associated software packages will be examined in Phase II for use with thermocouple flow sensors to determine air flow patterns.

A means for detecting air flow velocity and direction has many applications in addition to the radiation protection application for which the sensor is being developed in this project. This includes applications in both nuclear and non-nuclear facilities. A few examples of applications of low air flow sensors in nuclear and non-nuclear facilities are provided in Table 2.1.

TABLE 2.1

Typical Applications Of
New Flow Sensor

Nuclear Applications

- Fuel Fabrication Facilities
- Uranium Manufacturing Facilities
- Radiation Treatment Facilities
- Nuclear Waste Compacting and Handling
- Nuclear Power Plants
- Clean Up of Nuclear Weapons Production Sites
.
.
.

Non-Nuclear Applications

- Smoke Stack Air Monitoring
- Cooling Towers in Coal, Chemical, and Petroleum Plants
- Combustion Air Management for Boilers
- Gas Leak Tracing
- HVAC Duct Work, Exhaust, Vent Locations
- Environmental Air Monitoring
- Fire and Smoke Detectors Locations
.
.
.

---

3. AIR SAMPLING IN THE WORKPLACE

The information presented in this section is a summary of NUREG-1400, Regulatory Guide 8.25, and pertinent information from other sources. This section is included to provide a background on why measurement of low air flow rates and determining air flow patterns is important in radiation protection applications.

Air sampling in the workplace is performed to determine the quality of air inhaled by workers. The results of this sampling process can reveal information about the effectiveness of engineering design features including filtration and/or the confinement of harmful materials. In nuclear facilities, air sampling results are used to estimate worker radiation exposure in various areas of a plant to demonstrate compliance with regulatory dose limits and determine what protective measures are appropriate in those areas. Therefore, it is important for air samplers to provide accurate information about airborne concentrations of radioactive material in the area. Prompt and accurate detection of harmful contaminant levels is possible only with properly located sample points.

Proper evaluation of sampling information should include the identification of potential release points relative to worker locations, determination of air flow patterns, and a thorough knowledge of the location of air handling equipment in the area. The results of this evaluation can be used for the optimum placement of air sampling systems in radiological control areas.

3.1 Air Sampling Systems

Air sampling systems used in nuclear facilities usually consist of sample collection employing an appropriate collection medium (i.e. filters). These systems usually incorporate a vacuum pump and an air flow regulator to control the flow of air through the collector. The type of system used depends on the purpose of the air sampling, the type of airborne contaminant (particulate or gas), amount of detectable concentration and whether workers are temporarily present or present on a continuous basis.

Air samplers normally consist of a filter which is selected based on the physical and chemical properties of the airborne materials to be sampled, sampling efficiency, size, and resistance to air flow. The type of the filter used for radioactive particulates is different than those used for gases. Airborne radioactive particulates may be sampled with glass microfiber filters with various efficiency ratings or cellulose-ester-membrane filters with a wide range of compositions, pore sizes, collection efficiencies and flow resistances. On the other hand, radioactive gases like iodine, halogens and noble gases, and airborne tritium are sampled using chemically treated activated charcoal, desiccant or ionized chambers. The collection efficiencies of these media depends on the air flow rate, temperature, humidity, particle size and concentration. However, analysis of sampled radioactive halogens and noble gases with activated charcoal requires discrimination to properly measure iodine concentration. This is normally done by purging the charcoal after sampling to drive off the noble gases. In sampling for airborne tritium, the gaseous sample is normally passed over a catalyst such as palladium to convert it to tritiated water vapor. The tritiated water vapor is then passed through a water-filled bottle and the tritium is collected in the water. Desiccants such as silica gel, molecular sieves, anhydrous calcium sulfate or activated alumina can also be used. Real-time monitoring instruments also exist for direct measurement of tritium in air using an ionization chamber.[2]

Once the need for a specific sampling is identified, the appropriate air sampler is chosen to properly measure airborne radioactive material concentration. Four types of air samplers are typically used in nuclear facilities. These are lapel samplers, portable samplers, fixed-location samplers and general area radiation monitors. A brief description of these samplers is given in the following sections.

Lapel Samplers

Lapel samplers are personal air samplers which are worn on the upper part of a worker's protective clothing. These samplers are typically worn continuously while a worker is in an area with potential airborne contamination and are equipped with a battery-powered vacuum pump and a filter holder. The lapel samplers are among the more effective devices for estimating breathing zone concentration. A breathing zone is defined as the region within 30 centimeters (1 ft) of the worker's head or nose. Although these devices are widely used, they have some shortcomings. For example, their sampling is limited to 2 liters/minute which makes them unsuitable for areas with high airborne concentrations of radioactive material. Furthermore, they are expensive, noisy, bulky and uncomfortable to wear. They also can easily become contaminated and produce erroneous results for a worker's actual intake.

Portable Air Samplers

Portable air samplers are usually used in facilities where airborne concentrations of radioactive material change location frequently or are spread over a wide area. They also are used for special tasks and emergency situations where air sampling is needed for a short period of time. These samplers are equipped with a vacuum pump, a filter, and a rotameter, which is a volumetric flow rate measuring device. Their basic mode of operation is for air to be drawn into the sampler through a filter by means of a vacuum pump or a central vacuum system. The flow rate of air into the sampler is then measured by the rotameter. Portable air samplers are usually designed to accommodate both low and high volume air sampling. The advantage of these air monitors is that they can easily be carried to any location and be placed close to worker(s) being monitored. While not as accurate as lapel samplers, they can also be used for evaluating breathing zone exposure.

Fixed-Location Air Samplers

Stationary samplers are permanently installed at various locations where airborne contamination may be encountered and are called fixed-location air samplers. Using appropriate validation analyses, they may also be used for the evaluation of breathing zone airborne concentrations. These samplers are usually equipped with a filter and a vacuum pump or they can be connected to a central vacuum system. Air is continuously drawn into the sampler at a uniform rate through the filter. The filters are periodically collected and analyzed with radiation detection instruments to estimate average concentrations for the period of time sampled. Proper placement of these air samplers requires a knowledge of the air handling equipment, local air concentration data and typical worker locations. The results of fixed-location samplers and other samplers often vary significantly within the same area. For example, some studies have shown that the results of fixed-location samplers can vary significantly from those of lapel samplers.[2] This may be the case even if fixed-location samplers are properly positioned. The variation may be due to air flow pattern disturbances caused by workers moving, by operating equipment, or other factors which affect flow patterns.

General Area Radiation Monitoring Systems

Large area stationary monitors that are placed at several locations throughout a facility to provide early warning of increases in radiation counts are referred to as Continuous Air-Particulate Monitors (CAMs). A CAM has an alarm system to warn an operator whenever airborne activity exceeds a specified threshold. This system is usually equipped with stationary or moving filter tape which may be wrapped around a detector. This filter-detector unit is surrounded by a lead shield to block radiation from external sources. The newly designed CAMs utilize a different filter-detector design. These devices can provide rate counts in a very short time interval. They are normally used for qualitative monitoring of airborne concentrations and not for demonstrating regulatory compliance, or estimating breathing zone concentrations.

Concentrations of airborne materials are normally not uniformly distributed. In fact, they can vary widely within a given area. Improperly placed air samplers may not provide representative results for airborne concentrations. In such a case, therefore, the proper location of air samplers is of great importance in determining accurate concentrations. The air flow rate and patterns, as well as the particular objective of the air sampling are also important. For example, if the objective is to verify the effectiveness of containments, or to provide warnings about

increases in dose, sampling should be performed in the air flow pathway near the release point. On the other hand, if the objective of air sampling is to determine worker intake, the air sampler should be placed as close to the breathing zone of the worker as possible and practical. In general applications, an air sampler must be placed in the air flow pathway, downstream from the source of airborne radioactive material between the source and the worker. The identification of this flow pathway requires a thorough study of the air flow patterns under normal working conditions. Any changes in the work environment could bring about a change in the air flow pathway. This may necessitate reevaluation of the air flow patterns in the area. Air flow patterns are determined by the qualitative and quantitative methods discussed in the following sections.

3.2   Qualitative Air Flow Studies

Qualitative methods utilize visualization techniques to identify air flow direction and velocity and include the use of smoke candles, smoke tubes, helium-filled balloons and isostatic bubbles. When smoke is used, it is discharged at a location where there is a potential for airborne contamination in the work area. The air flow rate and direction can usually be determined by either visual observation, video tape, or photographs of smoke as it diffuses throughout the work area. Although this method is commonly used and is relatively easy and inexpensive to implement, the practice has certain limitations. Qualitative studies are most effective at low air velocities. The smoke diffuses too rapidly at high velocities (greater than 50 cm/sec) to allow tracking by observation. The smoke also leaves behind a residue which necessitates covering sensitive equipment in the area where the study is being performed. Furthermore, the smoke is a respiratory irritant which requires workers and test personnel to wear full-face respirators during its use. Proper lighting is also required in the monitoring area in order to visually track the smoke. This increased lighting could cause an elevation in the temperature in the area resulting in thermal currents that might skew the observations. In addition, the use of smoke in a nuclear facility requires coordination with fire protection because large amounts of smoke in an area can activate fire alarms.

The above concerns about the use of smoke for qualitative air flow studies require smoke tests to be performed when there is no work in progress or workers are not in the area. Therefore, the results of these studies may or may not represent the actual air flow patterns under normal working conditions. The impact could be improper placement of the air samplers in the area.

Helium-filled balloons or isostatic bubbles which float in the air with zero buoyancy are also used. The bubbles, which are filled with air and a small percentage of helium, can stay afloat for hours and are useful for detecting air flow direction, but they may adversely affect ongoing operations in the work area.

3.3   Quantitative Air Flow Studies

The quantitative method utilizes tracer gases and/or particle tracers with gas or particle detection devices to provide measurements of dilution effects in the work area. The method employs nonradioactive tracer aerosols at potential release points in the work area. The air movement and dispersion of aerosol, relative to the release point, can be identified using the ratio of concentrations at the detection point to concentrations at the release point. The method requires a large

array of detectors placed at different locations in the work area to identify and characterize air movements. Gaseous tracers do not address inertial effects caused by larger aerosol particles and they are not practical when filtered recirculation is used in a ventilating system. Although this method is effective, it is not in common use because it is expensive and very time consuming.

3.4   Conditions Which Affect Air Flow Patterns

There are many factors that can influence the air flow patterns in a work area. Placement of air samplers based only on the location of air supply and exhaust vents may not be appropriate. Placement based on smoke tests, which do not simulate the normal work environment, may also not be appropriate. In radiological facilities, air flow may have the following features:

- Cyclic Patterns: Due to poor ventilation systems, cyclic patterns can be formed which could cause the accumulation of contaminants or an increase in their concentration without it being detected by samplers at the exhaust ducts.

- Stagnation Patterns: Due to a combination of obstructions caused by the presence of large structures and/or equipment, location of the air supply and exhaust vents and high exhaust rates, streaming and/or stagnant layers can be formed in a work area which could cause elevated local concentrations of airborne radioactive material.

- Thermal Stratification: Heat generated by operating equipment, lights and workers can create thermal stratification within the work area. This stratification could lead to the development of thermal currents which force layers with higher temperatures to rise and cooler layers to fall. This condition may cause flows to come in from the bottom of a door or doorways and exit from the top.

- Bi-level Pattern: In the presence of poor ventilation systems, air can flow in opposite directions within the same area. These patterns generally exist in long narrow high ceiling areas such as hallways and operating galleries. When the air supply is located on one end of the area and the exhaust vent on the other end, two circulating air patterns may be formed. These type of flow patterns make it very difficult for fixed-location samplers to properly detect airborne concentrations of radioactive material.

- External Atmospheric Conditions: The outside atmospheric temperature, pressure and humidity may also affect the air flow patterns inside a facility. These effects are primarily encountered in large open areas with high ceilings such as a hot cell canyon area. The condition is more pronounced in facilities with high air infiltration through roll-up doors, poor air conditioning systems or poor insulation.

- Air Flow Modifiers: The activities that may alter air flow patterns are called air flow modifiers, such as:

  a. Opening doors, hoods, cells, etc.
  b. Moving and relocating equipment and large physical structures
  c. Movement and positioning of workers
  d. Changing HVAC operation modes
  e. Intermittently operated localized heater or air blowers.

These factors demonstrate the need for a device or devices that can identify air flow velocity and direction as well as temperature stratification in a work area. Such a device would make it possible to map this information into an area image representative of the air flow patterns in the area.

4. CURRENT TECHNOLOGIES FOR AIR FLOW MEASUREMENTS

In studies of gas and fluid flows, measurements of the speed and direction of the flow at specific points are usually needed to characterize the flow field. The flow field describes flow patterns in terms of the variation in flow velocity and direction from point to point. Various methods of flow visualization are available to provide a qualitative description of air flow patterns which may be sufficient in some cases; however, many applications require quantitative information.   Providing quantitative information requires accurate velocity measurements in several locations in order to properly characterize the flow field. Table 4.1 provides a listing of conventional flow sensors, their principle of operation and pertinent characteristics which include:[5]

    a) Sensitivity
    b) Rangeability (turndown ratio)
    c) Ease of installation and maintenance
    d) Repeatability

A few of the conventional flow sensors are described further below.

4.1 Pitot Tube

Pitot tubes operate on the principle of differential pressure. They are widely used in the measurement of air velocity for industrial and aerospace applications, aeronautical research, and in laboratory environments. The total pressure of a flowing air stream is the sum of the static pressure and the velocity pressure of the moving air. The Pitot tube is normally connected to a differential pressure measuring device, such as a manometer or a differential pressure transmitter, as shown in Figure 4.1. The tube senses the total pressure and the static pressure of the air stream,

therefore the output of the differential pressure measuring device is proportional to the velocity pressure and the air velocity of the flow stream.

For a steady one-dimensional air flow, the velocity is determined by the following Pitot equation:

$$V = \sqrt{(2 * (P_{stag} - P_{stat}) / \rho)} \quad (4.1)$$

where:

$$
\begin{aligned}
V &= \text{flow velocity} \\
\rho &= \text{fluid or gas density} \\
P_{stag} &= \text{free stream total pressure} \\
&\quad \text{or stagnation pressure} \\
P_{stat} &= \text{free stream static pressure}
\end{aligned}
$$

For the Pitot tube to accurately measure air velocity, its axis must be aligned with the flow stream, and the static taps on the tube's leading edge cannot be restricted. The Pitot tube is a very reliable and common device for air flow measurement at moderate flow rates. It is also inexpensive and easy to use, but is limited to measurement of moderate air flow rates since it loses sensitivity at low flow rates.

4.2 Hot Wire Anemometer

Hot wire anemometers are a common type of thermal mass flow measuring device. Typical applications include air velocity measurements for HVAC temperature and ventilation control, process gas measurements and control, combustion air measurements in boilers, environmental stack effluent monitoring, and air sampling for radiation protection applications.

Table 4.1
Conventional Instruments for Air Flow Measurements

Device	Principle of Operation	Characteristics
PITOT TUBE	Provides the kinematic difference between the static and the total pressure head of the system. The velocity is proportional to the square root of the differential pressure.	Inexpensive and easy to use. Unless individually calibrated, accuracy is only about +/- 5 percent. Minimum accurate measurable air flow is 3 m/s.
THERMAL METERS (Hot Wire Anemometer)	Hot wire anemometers measure fluid flow with output voltage proportional to the fluid velocity and the heat transfer characteristics of the fluid. The heat removed from the hot wire is proportional to the mass flow rate of the fluid.	The thermodynamic properties of the fluid will influence the heat transfer and response of the device. Primarily used in laboratory and research work with limited use in commercial applications.
LASER DOPPLER VELOCIMETER (LDV)	Uses the Doppler shift produced by the scattering of light by particles in a fluid flow stream. The Doppler shift frequency is proportional to the particle velocity.	The medium must be transparent, scattering particles are needed, and optical access is required for piping flows. System is expensive but very suitable for reference flow measurements in a laboratory. Primarily used for research.
DIFFERENTIAL PRESSURE SENSORS	Measures pressure drop across a flow element that is converted to flow rate. Three most commonly used flow elements are orifice, venturi, and nozzle. Primarily used for in-line piping flow.	Linear output range limited. Orifice meters are widely accepted, economical, but suffer from large pressure losses. Venturi meters have low pressure drop, but are expensive. Flow nozzles can be used at very high velocities, but have limited range.

	Table 4.1 (Continued)	
Device	**Principle of Operation**	**Characteristics**
VORTEX ANEMOMETER	Measures frequency of cyclic variations in fluid pressure induced by a bluff body placed in a flow stream.	Wide range with linear frequency output directly proportional to velocity. System costs are moderate.
TURBINE FLOW METERS	Flow passes through a rotor. The rotational speed is directly proportional to velocity.	The device has excellent accuracy over the operating range. It has linear real-time output which can be made directly available to an electronic monitoring system. Initial system costs are moderate but may become expensive to maintain.
SONIC VECTOR ANEMOMETER	Determines velocity by sensing the time of flight of ultrasonic sound pulses. Resolves magnitude and direction of air flow vectors at low air velocity.	Linear wide range output, simple clamp on installation for piping flows, and no pressure loss associated with other flow intrusive devices. Primarily used in research and laboratory applications.
FLOW VISUALIZATION TECHNIQUES	Uses time lapse strobe photography, or high speed movies with smoke injections, pulsed smoke injections, or particle scattering.	Primarily a qualitative technique for studying air flow and dispersion patterns.

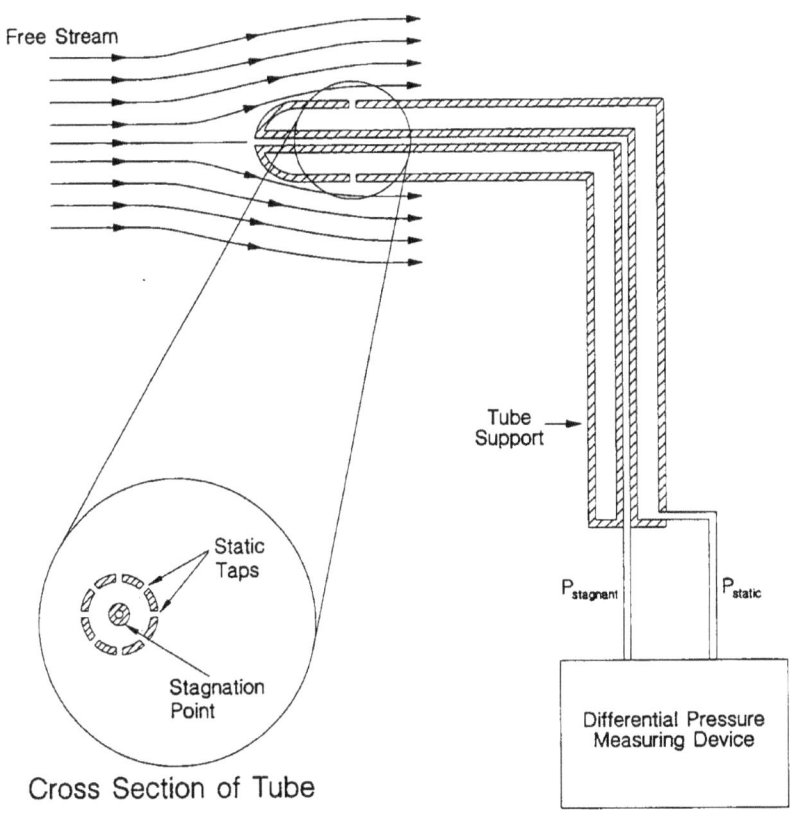

Free Stream

Tube
Support

$P_{stagnant}$

P_{static}

Static
Taps

Stagnation
Point

Cross Section of Tube

Differential Pressure
Measuring Device

Figure 4.1 Typical Pitot Tube Configuration

A hot wire anemometer was used in the early development and testing phase of this project. The device performance was excellent for measuring air velocities as low as 50 cm/sec; however, it demonstrated a loss of sensitivity at velocities less than 50 cm/sec and a minimum threshold detection of approximately 6 cm/sec. The device also had slow response to changes in velocity.

Two basic types of hot wire anemometers are in common use. These are the Constant Power Anemometer (CPA) and the Constant Temperature Anemometer (CTA). For the CPA, a resistive sensor element is heated by a constant power input. The temperatures of the heated element and the ambient flow are measured by changes in the resistance of two Resistance Temperature Detectors (RTDs). Platinum wire elements are typically used in these RTDs to provide linear and repeatable resistance measurements proportional to temperature. The differential temperature between the heated sensor element and the ambient temperature sensor element is inversely proportional to the flow rate of the fluid. Any change in the flow velocity results in a change in the differential temperature, which is detected as a change in resistances of the two RTD elements. Signal conditioning and processing circuitry converts the resistance changes to a voltage signal output which is proportional to the fluid velocity. A typical hot-wire anemometer is shown in Figure 4.2.

Constant power anemometers have a slow response to changes in fluid flow and ambient temperature conditions. Due to problems caused by high operating temperatures at very low flows, CPAs do not have a true zero reading at zero flow. Constant temperature anemometers correct some of these limitations. A feedback control system maintains a constant differential temperature between the two RTD sensing elements by varying the

power input to the heated element. The flow velocity can be correlated to the power input. The temperature of the heated element is much lower than that of a comparable CPA, therefore, the problems with free convection currents are reduced. A temperature compensation feedback system may also be used to improve the speed of response of the device.[6]

The CTA is one of the more recent advancements in hot wire flow measurement techniques. It has primarily been used in research applications with limited use in the industrial market. Although it is an excellent device for many applications, it is still constrained by a loss of sensitivity at low air flow rates. Nearly all hot wire anemometers have diminishing accuracy below 40 cm/sec and a lower velocity measuring limit of approximately 10 cm/sec.

4.3   Laser Doppler Velocimeter

The Laser Doppler Velocimeter (LDV) is a very accurate instrument for local measurement of gas and fluid velocities. It has several advantages over the Pitot tube and the hot wire anemometer, including the following:

1) Measurement of velocity is direct rather than inferred from physical properties such as pressure as in the Pitot tube or by heat transfer coefficient as in the hot-wire anemometer.

2) No probe needs to be inserted into the flow for velocity measurement (i.e., the process is not disturbed).

3) It has a fast response time (almost instantaneous).

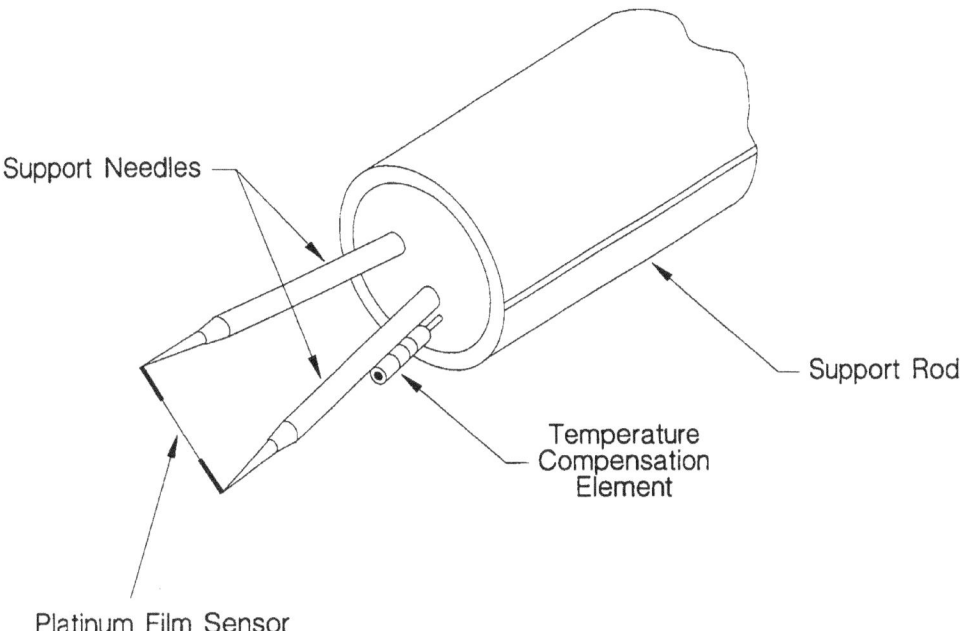

Support Needles

Support Rod

Temperature
Compensation
Element

Platinum Film Sensor

Figure 4.2 Typical Hot-Wire Probe

LDVs are used for a variety of research and industrial process flow applications; however, they do have some limitations. Use of LDVs requires a transparent flow medium, tracer particles in the fluid, and optical access for piping flows. Due to system complexity, setup, and operating costs, LDVs are still primarily specialized research tools for laboratory use.

The LDV operates by measuring the Doppler shift produced by the scattering of light as a result of the movement of particles in a fluid flow stream. A typical configuration for a LDV system is illustrated in Figure 4.3. As shown in this figure, a coherent monochromatic light is emitted from an air or water cooled laser source, typically argon-ion. The laser light is separated into two beams of equal intensity by the beam splitter and is coupled via fiber optics to the dual transmitter/receiver probe. The beams are projected and focused into the fluid flow stream where the crossing point of the two beams defines a measuring volume for determining flow velocity. Figure 4.4 shows the beam projection and crossing point for measurement of air flow velocity. The measuring volume consists of a fringe interference pattern formed by the constructive summation of the incident light waves. The distance between the fringes (d_f) is determined by the probe optics and is given

$$d_f = \lambda / 2 \sin (\alpha / 2) \qquad (4.2)$$

where λ is the wavelength of the beam and α is the angle between the two incident beams. Small particles are released into the flow and scatter the light with a frequency shift proportional to the particle speed and the fringe spacing. The scattered light is detected by the probe and coupled by fiber optics to photomultiplier detectors in the receiver where they are converted to electrical signals. The electrical signals are then sent to a signal processor where the Doppler shift frequency is determined. The velocity of the particles (V) and the fluid flow is calculated as a function of the Doppler shift frequency (f) and the fringe spacing using Equation 4.3.[7]

$$V = f \, d_f = f \lambda / 2 \sin (\alpha / 2) \qquad (4.3)$$

The LDV data acquisition and analysis system acquires velocity measurement data, typically in blocks of 1024 samples. Data processing programs calculate statistical parameters such as the mean velocity, variance, and the standard deviation of the data distribution. The quality of the data is evaluated by parameters associated with a normal distribution function. An LDV data processor is shown in Figure 4.5, and a typical data distribution is shown in Figure 4.6. The Laser Doppler velocity measurement range is virtually unlimited with accuracy within 0.1% of reading and it requires no calibration.

4.4   Other Flow Measurement Devices

Several other conventional flow measurement devices were surveyed to determine their suitability for use in the calibration of the new thermocouple air flow sensor. The devices studied were differential pressure sensors, laminar flow elements, vortex anemometers, and turbine flow meters. The characteristics, advantages and limitations of each are described below.

Differential pressure sensors use the change in fluid velocity and pressure along with a change in flow area to measure the flow rate. Three major types of differential pressure sensors are orifice, venturi, and nozzle devices. Differential pressure devices are widely accepted and typically used to measure in-line process fluid flows; however, they have some

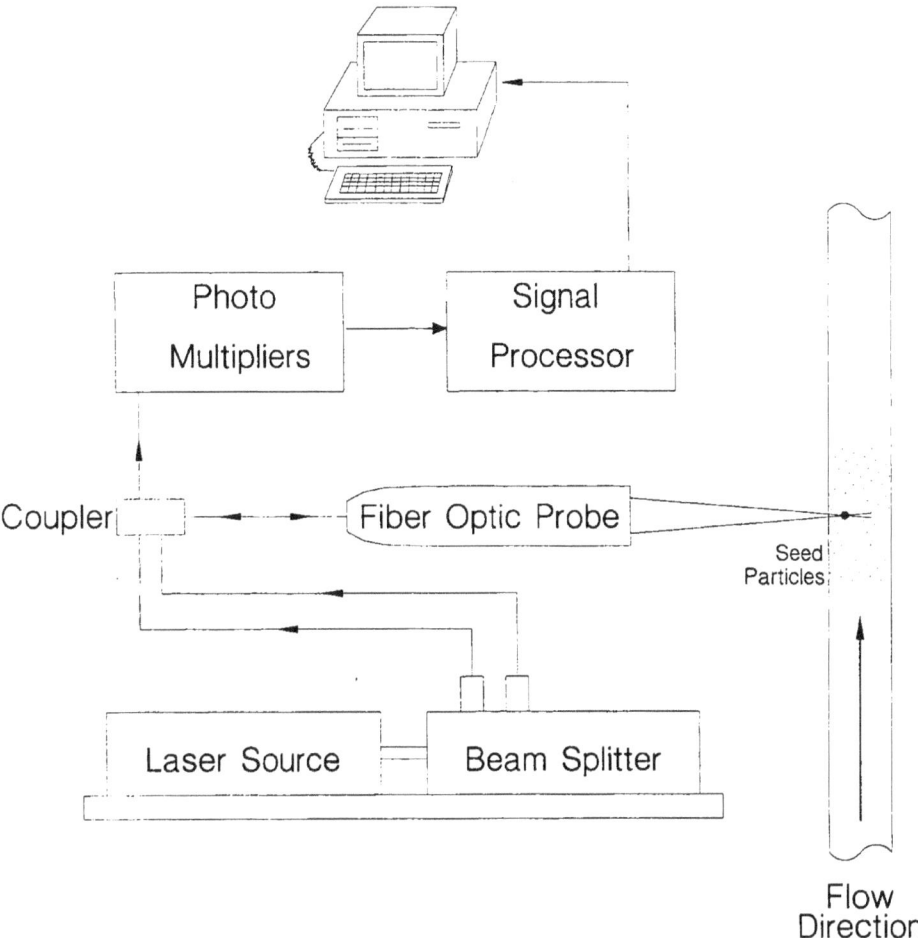

Figure 4.3 Typical LDV System Configuration

Figure 4.4 LDV Measuring Air Flow

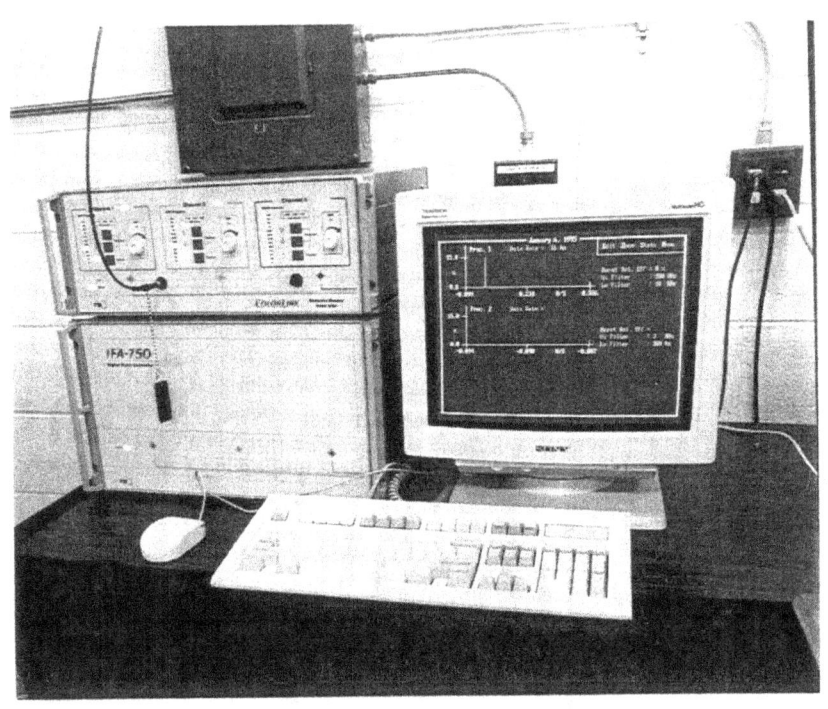

Figure 4.5 LDV Data Processing Equipment

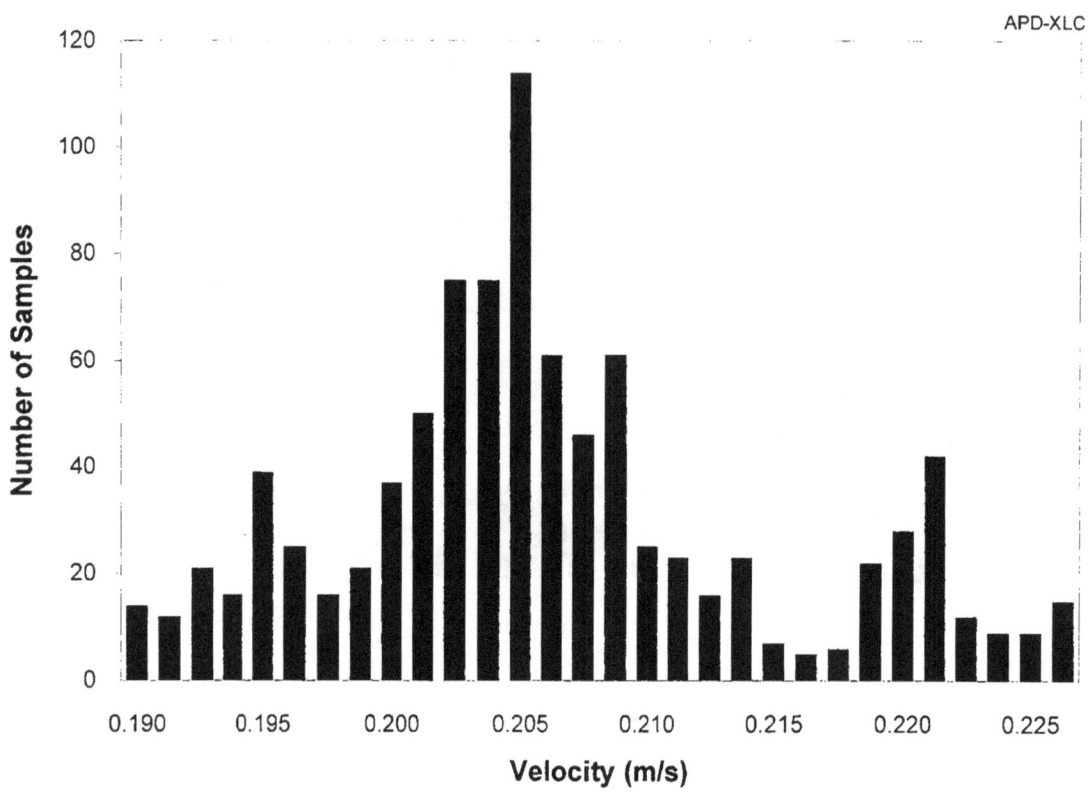

Figure 4.6 Typical Air Velocity Data Distribution Plot Using An LDV

disadvantages. A sufficient pressure drop must exist to obtain a valid flow measurement. Errors increase at low flows, and the linear output range is limited. Rangeability, which is the ratio of maximum to minimum accurate flow measurement, is typically only three to one. Although differential pressure devices are suitable for measuring medium to high velocity air flows in piping systems, they cannot be used to measure typical ambient air flows.[8]

The Laminar Flow Element (LFE) works on the principle of volumetric flow through capillary tubes. The LFE produces a differential pressure output which is linearly proportional to the volumetric flow through the device. The LFE may be used for calibration of air and gas flow sensors because of its high accuracy, excellent repeatability, and wide rangeability. Accuracy is within 0.25% of reading, repeatability is within 0.1%, and turndown ratio is ten to one. However, ambient air flows do not produce sufficient differential pressure across the LFE to be accurately measured.

The vortex anemometer utilizes the phenomena of vortex shedding. Eddy currents are generated by turbulence induced by a bluff body placed in a flow stream. The eddies form a vortex with a frequency corresponding to cyclic variations in the fluid pressure. This frequency is directly proportional to the fluid velocity and can be detected by hot wire, fiber optic, piezo-electric, ultrasonic, or other techniques. When used as a volumetric flow meter, the device must be placed with a specified length of straight line pipe upstream and downstream of the device. If used as a wind speed sensor, the device is placed in a free flow stream. The vortex anemometer has

the advantage of having a wide range linear frequency output proportional to air velocity over a range of 1 to 60 m/s; however, this is not sufficient for measurement of air flows typical of radiation protection applications.[9]

Turbine flow meters, such as propeller anemometers, are precision velocity measuring devices. Air flow passes through a free-turning rotor with the rotational speed directly proportional to flow rate, except at low velocities. The non-proportional region at low flows is due to initial rotor inertia and bearing friction. The relationship between flow and rotation depends on the design of the rotor blades. The device has excellent accuracy over the full operating range, and a linear real time output which can be made directly available to an electronic monitoring system. Rangeability is excellent at approximately 100/1 with a minimum threshold sensitivity of approximately 30 cm/sec. Ambient air flow movements are generally less than this which precludes the use of this device for health physics and radiation protection applications.

In order to accurately calibrate and determine the feasibility of the new thermocouple air flow sensor, a very accurate and reliable standard device for low air velocity measurements had to be selected. The LDV was identified as having the best accuracy for measurement of air flow velocities ranging from near stagnation to highly turbulent flow regimes. Its accuracy is independent of process conditions and the results are repeatable and reliable. These features made it a very suitable device for calibrating the new thermocouple air flow sensor developed during this project.

5. THERMOCOUPLE RESPONSE TIME VERSUS FLOW RATE

This section provides the derivation of a general correlation between response time of a temperature sensor and the fluid flow rate to which the sensor is exposed. The correlation is used in the development of thermocouple flow sensors to convert response time information to flow rate information.

5.1 Technical Background

The response time of a thermocouple consists of an internal component and a surface component. The internal component depends predominately on the thermal conductivity (k) of materials inside the thermocouple, while the surface component depends on the film heat transfer coefficient (h). The internal component is independent of the process conditions except for the effect of temperature on material properties inside the thermocouple. The surface component is predominately dependent on process conditions such as flow rate, temperature, and to a lesser extent, the process pressure. These parameters affect the film heat transfer coefficient which increases as process parameters such as flow rate and temperature are increased. Figure 5.1 illustrates how the response time of a thermocouple will decrease as (h) is increased. In this illustration, the effect of temperature on material properties inside the sensor is neglected.

In the study of process effects on response time, another factor that should be considered is the ratio of internal heat transfer resistance to the surface heat transfer resistance. This ratio is called the Biot Modulus (N_{Bi}) which is given by:

$$N_{Bi} = \frac{\text{internal heat transfer resistance}}{\text{surface heat transfer resistance}} = \frac{hr_o}{k}$$

If the Biot Modulus is large, then the response time may change very little as h is increased. If the Biot Modulus is small, the response time will be very sensitive to changes in h, especially in poor heat transfer media where h is small. Figure 5.2 shows the response time of two sensors in room temperature water as a function of flow rate. One of the sensors was tested inside a thermowell and the other was tested without a thermowell. It is apparent that the response time is not as sensitive with respect to flow rate for the sensor with the thermowell, as compared to the sensor without a thermowell. This is because the internal resistance of the sensor-thermowell combination dominates its surface resistance, while the internal and surface resistances of the sensor without the thermowell are closer to one another.

5.2 Response Time Versus Heat Transfer Coefficient

As shown in Figure 5.1, the response time of a thermocouple decreases as the heat transfer coefficient is increased. In order to derive the correlation between the heat transfer coefficient and response time, we should note that the time constant (τ) or response time of a thermocouple may be written as:

$$\tau = \frac{mc}{UA} \qquad (5.1)$$

where m and c are the mass and specific heat capacity of the sensing portion, and U and A

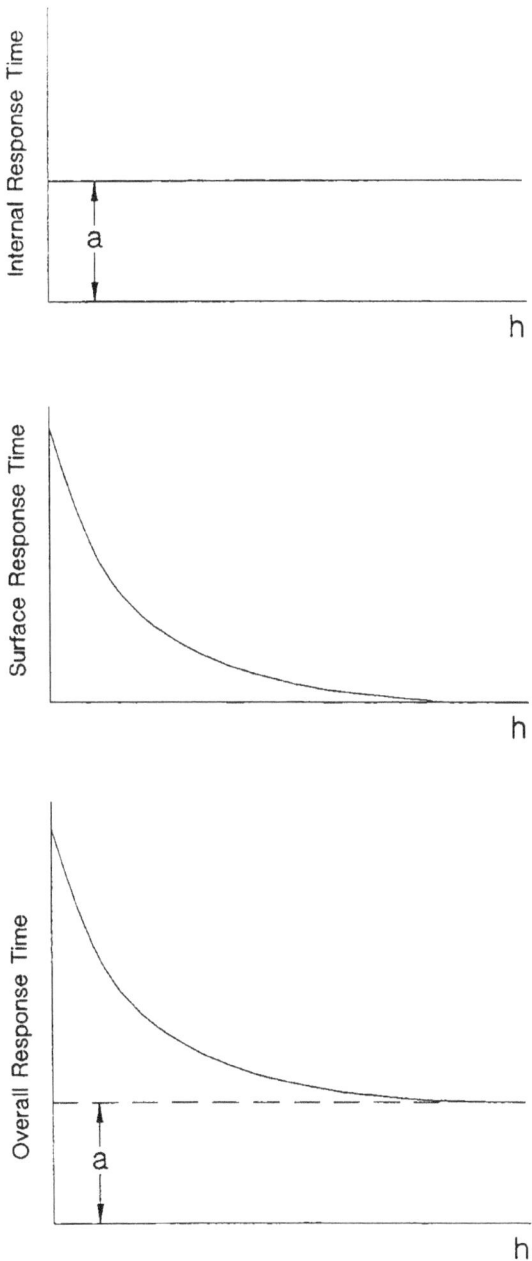

Figure 5.1   Changes in Internal and Surface Components of Response
Time as a Function of Heat Transfer Coefficient.

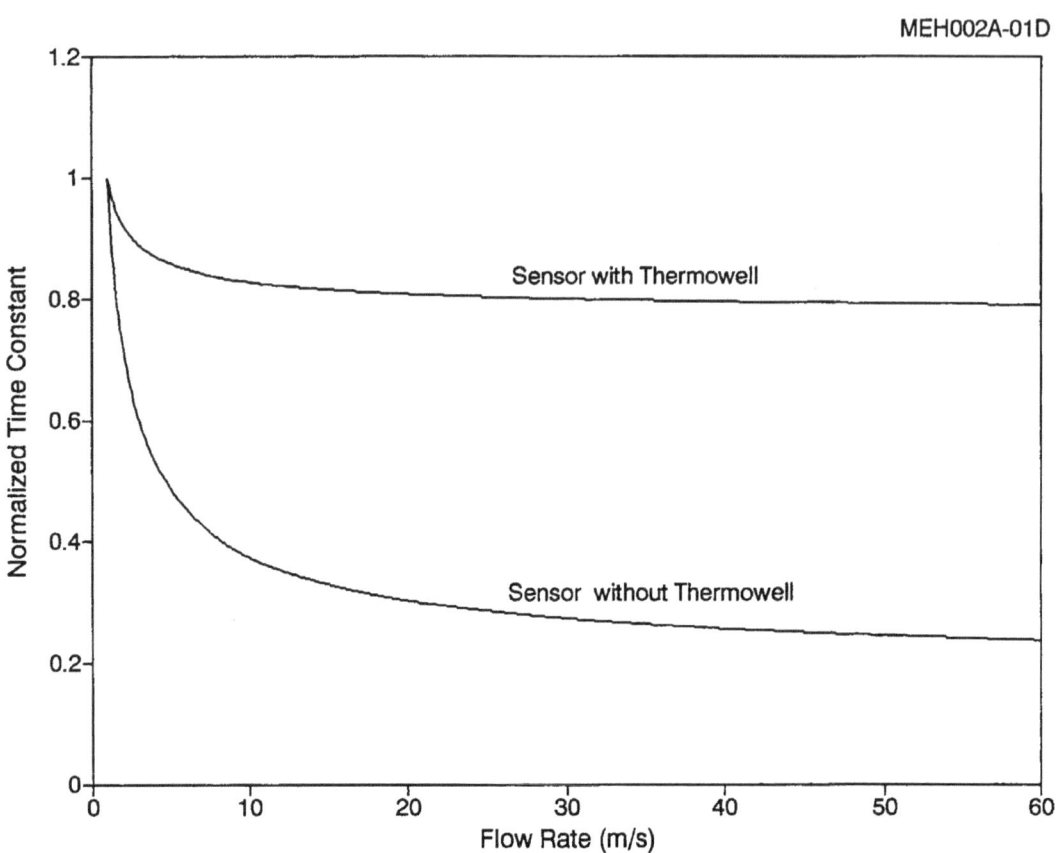

Figure 5.2   Response-Versus-Flow Behavior of a Sensor
Tested With and Without Its Thermowell.

are the overall heat transfer coefficient and the affected surface area of the thermocouple. Note that the overall heat transfer coefficient U is used, as opposed to the film heat transfer coefficient, h, accounting for both internal and external heat transfer resistances. The overall heat transfer coefficient accounts for the heat transfer resistance both inside the sensor and at the sensor surface. More specifically,

$$UA = \frac{1}{R_{tot}} = \frac{1}{R_{int} + R_{surf}} \qquad (5.2)$$

where :

R_{tot} = total heat transfer resistance
R_{int} = internal heat transfer resistance
R_{surf} = surface heat transfer resistance.

For a homogeneous cylindrical sheath, the internal and surface heat transfer resistances may be written as follows for a single-section lumped model:[10]

$$R_{int} = \frac{\ln (r_o/r_i)}{2\pi kL} \qquad (5.3)$$

$$R_{surf} = \frac{1}{2\pi hLr_o} \qquad (5.4)$$

where :

r_o = outside radius of thermocouple
r_i = radius at which the junction is located
k = thermal conductivity of sensor material
L = effective heat transfer length
h = film heat transfer coefficient.

Substituting Equation 5.3 and 5.4 in Equation 5.1 and 5.2 yields :

$$\tau = \frac{mc}{UA}$$

$$= mc \left[\frac{\ln(r_o/r_i)}{2\pi kL} + \frac{1}{2\pi hLr_o} \right] \qquad (5.5)$$

Since $m = \rho \pi r_o^2 L$ we can write :

$$\tau = \frac{\rho c r_o^2}{2k} \left[\ln (r_o/r_i) + \frac{k}{hr_o} \right] \qquad (5.6)$$

where ρ is the density of the material in the sensor. Note that the second term in Equation 5.6 is the reciprocal of the Biot Modulus.

Equation 5.6 may then be re-written in terms of two constants C_1 and C_2:

$$\tau = C_1 + C_2 / h \qquad (5.7)$$

where :

$$C_1 = \frac{\rho c r_o^2}{2k} \ln (r_o/r_i) \qquad (5.8)$$

$$C_2 = \frac{\rho c r_o}{2} \qquad (5.9)$$

Equation 5.7 can be used to estimate the response time of a thermocouple installed in a process based on response measurements made previously in a laboratory. Laboratory response time measurements may be made in at least two different heat transfer media (with different values of h) to identify C_1 and C_2. Once C_1 and C_2 are identified, Equation 5.7 can be used to estimate the response time of the thermocouple in the actual process media. The value of h for this calculation can be estimated from characteristics of the media and the process temperature, pressure, and flow conditions.

5.3 Response Time Versus Flow Correlation

A correlation for response time versus fluid flow rate may be derived from the relationship between the heat transfer coefficient (h) in Equation 5.7 and fluid flow rate (u).

The heat transfer coefficient is obtained using general heat transfer correlations involving the Reynolds, Prandtl, and Nusselt numbers which have the following relationship:

$$Nu = f(Re, Pr) \qquad (5.10)$$

In this equation, $Nu = hD/K$ is the Nusselt number, $Re = Du\rho/\mu$ is the Reynolds number, and $Pr = C\mu/K$ is the Prandtl number. These numbers are all dimensionless and their parameters are defined as follows :

h = film heat transfer coefficient
D = sensor diameter
K = thermal conductivity of process fluid
u = average velocity of process fluid
ρ = density of process fluid
μ = viscosity of process fluid
C = specific heat capacity of process fluid.

For the correlation of Equation 5.10, several options are available in the literature for flow past a signal cylinder. One of the common correlations is that of Rohsenow & Choi,[11] and the other is from Hilpert.[12] The Rohsenow & Choi correlation is:

$$Nu = 0.26\ Re^{0.6}\ Pr^{0.3} \qquad (5.11)$$
$$for\ \ 1,000 < Re < 50,000$$

and the Hilpert general correlation is:

$$Nu = C\ Re^{n}\ Pr^{1/3} \qquad (5.12)$$

where:

$C = 0.989$ $n = 0.330$ for $0.4 < Re < 4$
$C = 0.911$ $n = 0.385$ for $4\ \ < Re < 40$
$C = 0.683$ $n = 0.466$ for $40\ < Re < 4000$.

The second correlation covers a more specific range of Reynolds numbers and is more suited for air flowing at very low to moderate velocities, while the first correlation is more suited for water. Substituting Equation 5.11 or 5.12 in Equation 5.10 will yield :

$$h = C_1'\ u^{0.6}\ \ or\ \ h = C_2'\ u^{n} \qquad (5.13)$$

where C_1' and C_2' are constants and u is the fluid flow rate. Substituting the relations given by 5.13 in Equation 5.7, we will obtain the correlations between the response time and fluid flow rate:

$$\tau = C_1 + C_3\ u^{-0.6} \qquad (5.14)$$

or

$$\tau = C_1 + C_4\ u^{-n} \qquad (5.15)$$

where the value of n is given in Equation 5.12.

Either one of the above two equations may be used to estimate the response time of a thermocouple as a function of flow rate. In this project, Equation 5.15 is used in the response versus flow experiments as described later in this report.

With either of the two Equations 5.14 or 5.15, measurements at two or more flow rates in water or air in a laboratory may be made and the two constants of the response versus flow correlation for the thermocouple of interest may be identified. Once these constants are identified, they can be used to estimate the response time of the thermocouple in other media for which the flow rate (u) is known, and/or estimate the flow rate if the response time (τ) is known. Note that constants for Equation 5.15 vary with the Reynolds number. Therefore, three separate correlations are needed to cover the appropriate ranges of air velocity.

As a part of this project, a series equation was developed based on experimental data acquired in the AMS laboratory characterizing response times and air flow rates for Reynolds numbers from 0.2 to 800. The significance of the series equation is that one equation is suitable for all three ranges of Reynolds numbers described in Equation 5.12, as opposed to three separate equations. The details of the series equation is given later in this report.

5.4 General Effects of Temperature on Response Time

Unlike flow, the effect of temperature on response time of a thermocouple may not be estimated with great confidence. This is because temperature can either increase or decrease the response time of a thermocouple. Temperature affects both the internal and surface components of the response time. Its effect on the surface component is similar to that of flow. That is, as temperature is increased, the film heat transfer coefficient (h) generally increases and causes the surface component of response time to decrease. However, the effect of temperature on the internal component of response time is more subtle. High temperatures can cause the internal component of response time to either increase or decrease depending on how temperature may affect the properties as well as the geometry of the material inside the

thermocouple. Due to differences in the thermal coefficient of expansion of materials inside the thermocouple and the sheath, the insulation material inside the thermocouple may become either more or less compact at higher temperatures. Consequently, the thermal conductivity of the thermocouple material and therefore the internal response time can either increase or decrease. Furthermore, voids such as gaps and cracks in the thermocouple construction material can either expand or contract at high temperatures and cause the internal response time to either increase or decrease depending on the size, the orientation, and the location of such voids. At high temperatures, the sheath sometimes expands so much that an air gap is created at the interface between the sheath and the insulation material inside the thermocouple. In this case, the response time can increase significantly with temperature.

In experiments conducted by Carroll and Shepard[13] in a Sodium loop at the Oak Ridge National Laboratory (ORNL), more than a dozen insulated junction Type K sheathed thermocouples with Magnesium Oxide (MgO) insulation were tested for the effect of temperature on response time. All of these thermocouples were found to have a larger response time at higher temperatures. Figure 5.3 shows two examples of the ORNL results. The thermocouples in the ORNL experiments were all 0.16 cm in outside diameter and were tested in flowing Sodium at approximately 0.6 cm per second. It was further determined by ORNL that the effect of temperature on response time of different thermocouples is different. The effect of temperature on an identical group of thermocouples was confirmed to differ from one thermocouple to another. Therefore, a general response time versus temperature relationship could not be determined for the thermocouples tested by ORNL.

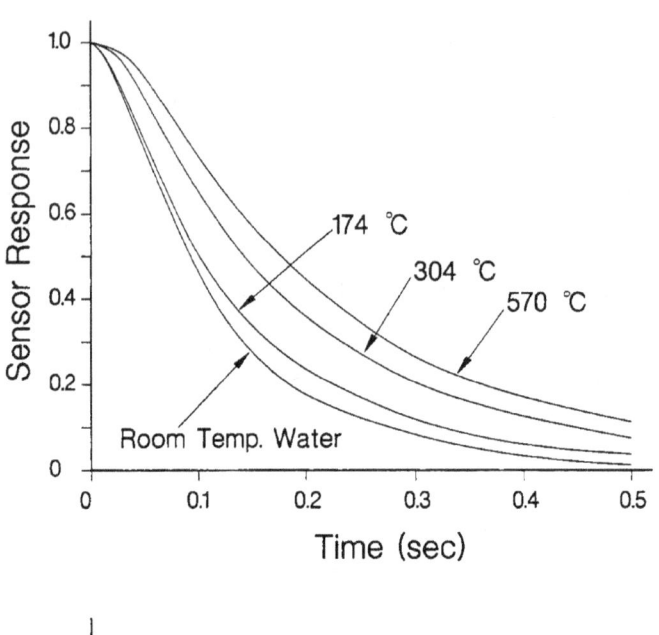

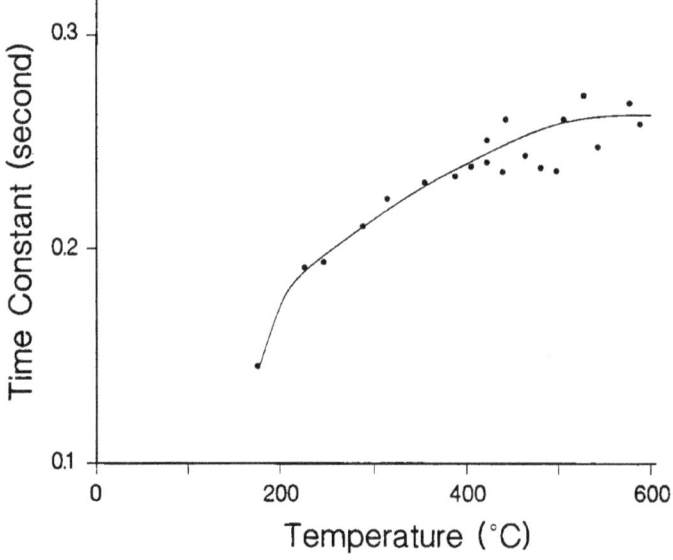

Figure 5.3   Examples of Effect of Temperature on Response
Time of Sheathed Thermocouples.

6. RESPONSE TIME TESTING TECHNIQUES

The thermocouple flow sensor is based on measuring the response time of a thermocouple and converting the results to a flow signal using a pre-determined response time-versus-flow correlation. These correlations are generated in a laboratory using the plunge test method or the LCSR method, whichever is more convenient for the laboratory work. The plunge or LCSR tests are performed at several widely spaced known flow rates and the results used with equations 5.7, 5.14, or 5.15 to identify the constant of the equations so that the equation can be used to convert response time results to corresponding flow rates.

The plunge and LCSR methods are described in this chapter.

6.1 Plunge Test

The response time of a thermocouple is classically measured in a laboratory environment using a method called the plunge test. In this test, the thermocouple is exposed to a sudden change in temperature and its output is recorded until it reaches a steady state value. The analysis of a plunge test to obtain the time constant of a thermocouple is simple. For example, if the thermocouple output transient is recorded on a strip chart recorder, the time constant is found by measuring the time necessary for the sensor output to reach a value corresponding to 63.2 percent of the final value (Figure 6.1). It should be noted that although this definition of time constant is analytically valid for only a first order system, it is used conventionally for determining the response time of all temperature sensors regardless of the dynamic order. All references to the terms

response time or time constant in this report correspond to this definition regardless of the type or size of the thermocouple, the test condition, or the test method being used.

The step change in temperature that is needed for response time testing is usually produced by plunging the thermocouple from a media at a given temperature into another media at a different temperature. The test is normally conducted in either water or air.

Testing in water may be accomplished through a number of methods. One method is to allow the thermocouple to reach thermal equilibrium in room temperature air and then plunge it suddenly into warmer or cooler water. The temperature of the final media, in this case the water, must be held constant during the test. A similar method is to heat the thermocouple in air above the water using a warm air blower and then plunge it into room temperature water. The same procedures are used for testing of thermocouples in air. In this project, the tests that were performed in air involved heating the thermocouple with a warm air blower and plunging it into an air stream at ambient temperature. Figure 6.2 shows simplified schematics of the test equipment used to perform plunge testing in water and air.

The thermocouple response time obtained by the plunge method is a relative index which should be accompanied by an expression of the test conditions. This is important because the response time of a thermocouple is strongly dependent on the properties of the final media in which it is plunged. The type of media (air, water, etc.) and its flow rate and temperature must always be mentioned with the response time results. The flow rate is usually the most

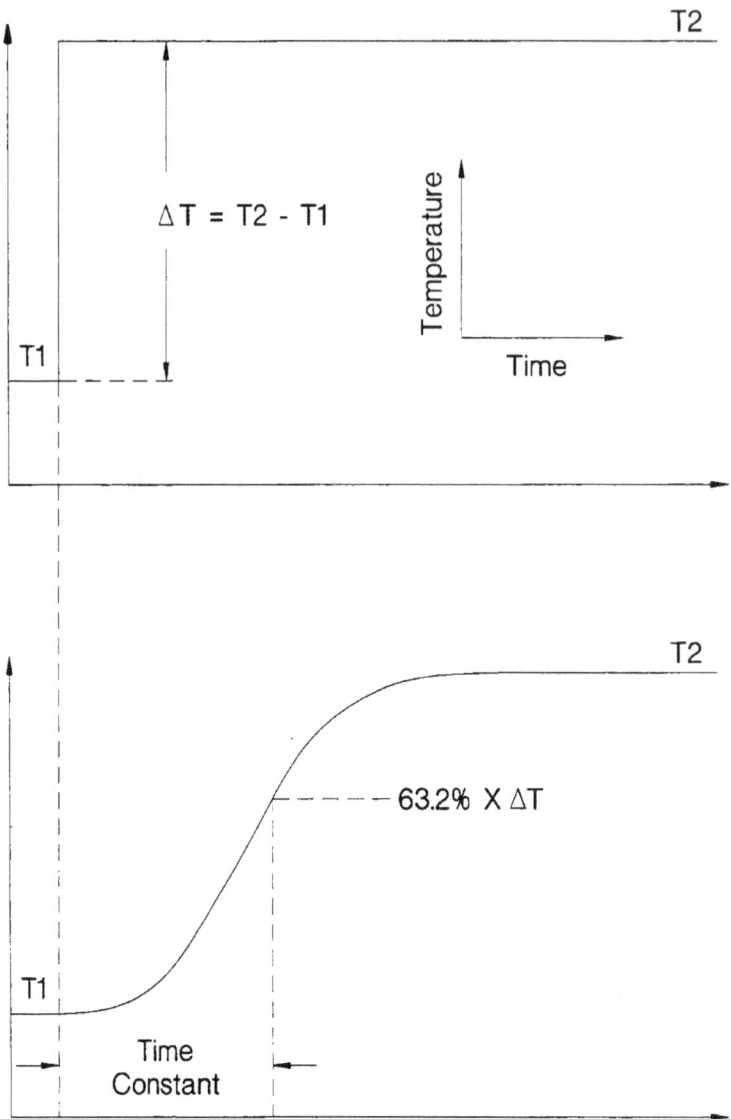

Figure 6.1   Determination of Thermocouple Response Time
              from a Plunge Test Transient.

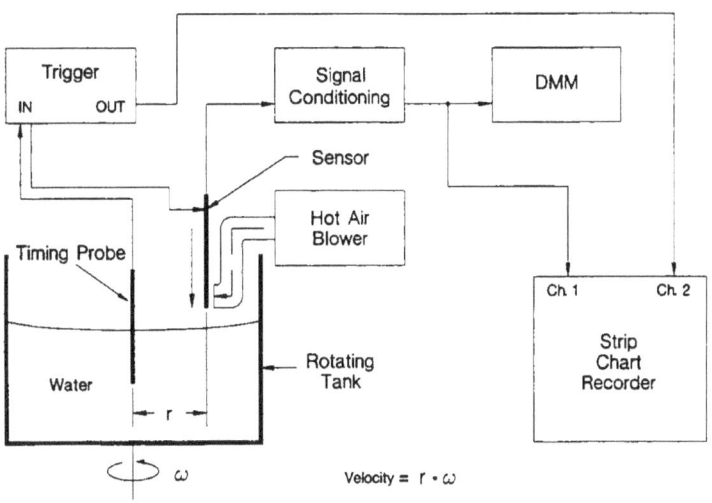

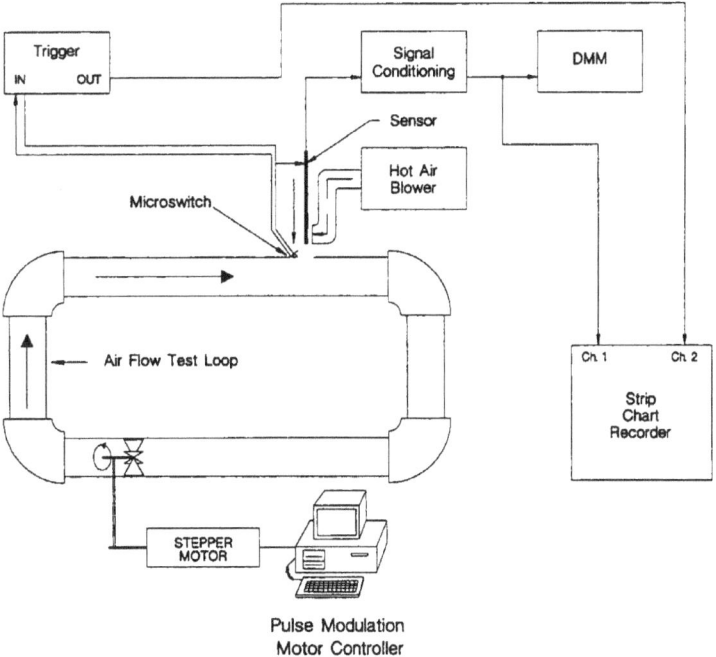

Figure 6.2   Equipment Setup for Laboratory Plunge
Tests in Water and Air.

important factor followed by temperature. The significance of these process parameters was described in an earlier section.

6.2 Loop Current Step Response Test

Since the response time of a thermocouple is strongly affected by process conditions, laboratory measurements such as plunge tests at reference conditions cannot provide accurate information about the "in-service" response time of a thermocouple. Therefore, an in-situ method that can be implemented at process operating conditions must be used. The Loop Current Step Response (LCSR) test method was developed to provide the in-situ response time testing capability. The test is performed by heating the thermocouple internally by applying an electric current to its extension leads (Figure 6.3). The current is applied for a few seconds to raise the temperature of the thermocouple a few degrees above the ambient temperature. The current is then cut off and the thermocouple output is recorded as it cools to the ambient temperature (Figure 6.4). This transient, which is referred to as the LCSR transient, is primarily due to the cooling of the thermocouple junction. The time necessary for the thermocouple to cool to steady-state is proportional to its ability to dissipate the heat generated in its junction. Therefore, the LCSR transient data may be used with an analytical approach to identify the response time of the thermocouple under the conditions tested. The analytical approach uses the LCSR data to establish the response of the sensor to any change in temperature. The validity of the LCSR test can be demonstrated by measuring response times of a group of thermocouples by the plunge method in a laboratory and repeating the measurements in the same conditions using the LCSR method.

The validation work performed in this project is described in Section 8 of this report entitled, "LCSR Validation Tests."

Figure 6.5 shows typical LCSR transients for thermocouples of different sizes. The transients have been inverted and normalized to start at zero and proceed in an increasing direction although the LCSR cooling transients for thermocouples start high and proceed low. Note in Figure 6.5 that the rate of change of LCSR transient (i.e., the dynamic response characteristics of these thermocouples) is proportional to the thermocouple size. Figure 6.6 shows the three transients on the same plot. The thermocouple sizes in these figures are given in mils where one mil is 1/1000 of an inch.

The LCSR testing of thermocouples can be performed using an AC or a DC current source to produce Joule heating, which is proportional to the current squared and is distributed along the entire length of the thermocouple. The Joule heating is given by I^2R where I is the applied current and R is the electrical resistance of the thermocouple wires involved. Since the electrical resistance of a thermocouple circuit is small and distributed along the entire length of the sensor, the heating current must be large enough to produce sufficient heating and provide a useful LCSR signal when the current is cut off. Depending on the size and length of the thermocouple and its extension wires, heating currents between approximately 0.2 to 2.0 amperes are typically used in LCSR testing of thermocouples. In addition to Joule heating, the application of an electric current to a thermocouple produces Peltier heating or cooling depending on the direction of the applied current. The Peltier effect can cause a problem if DC currents are used in LCSR testing of thermocouples. While Joule heating is uniformly distributed along the

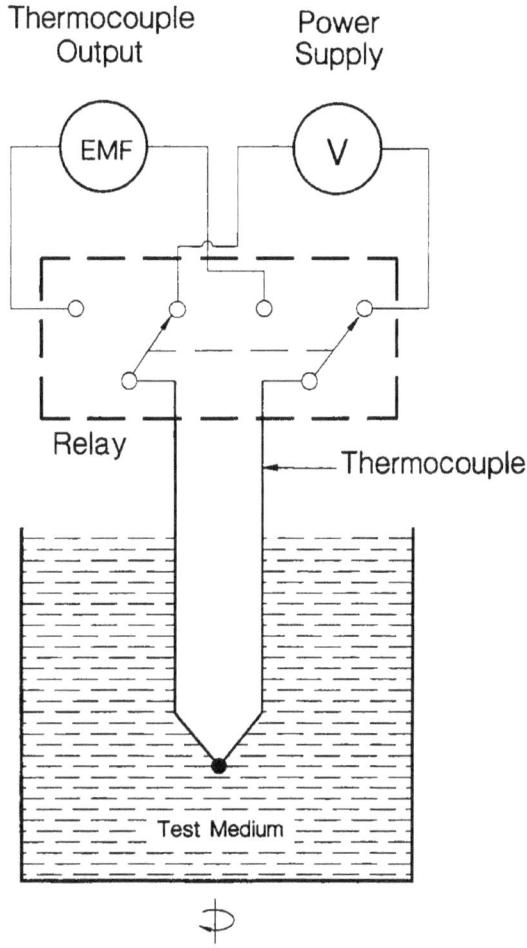

Figure 6.3 Simplified Schematic of LCSR Test Equipment.

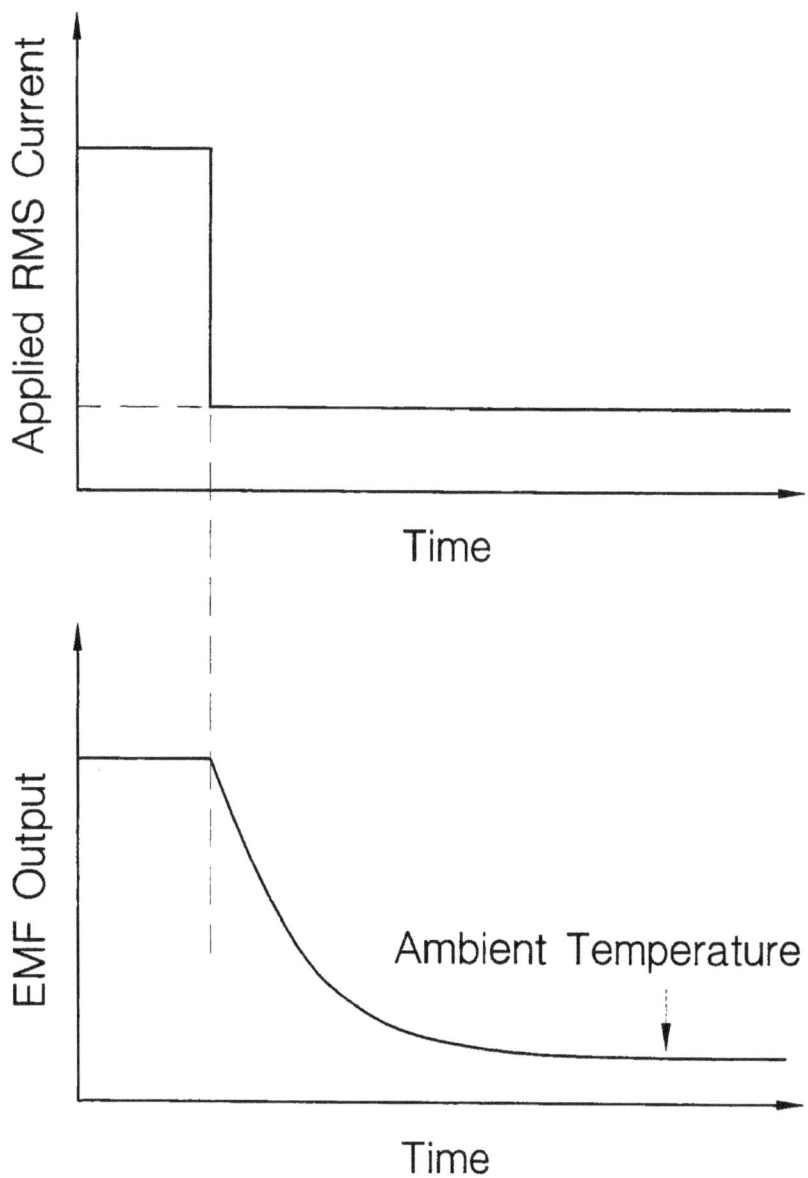

Figure 6.4 A Typical LCSR Cooling Transient.

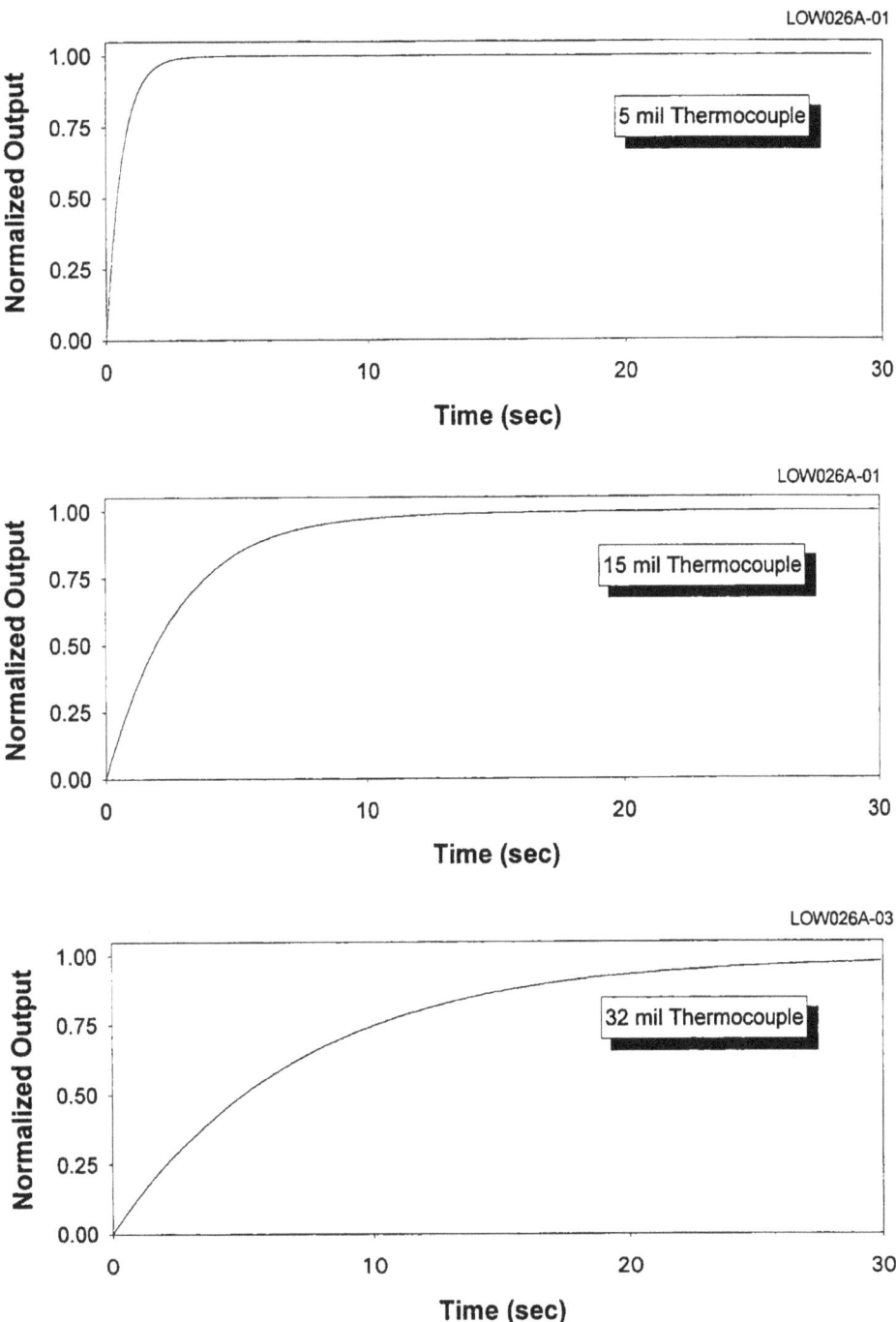

Figure 6.5 Typical LCSR Transients For Different Size Thermocouples

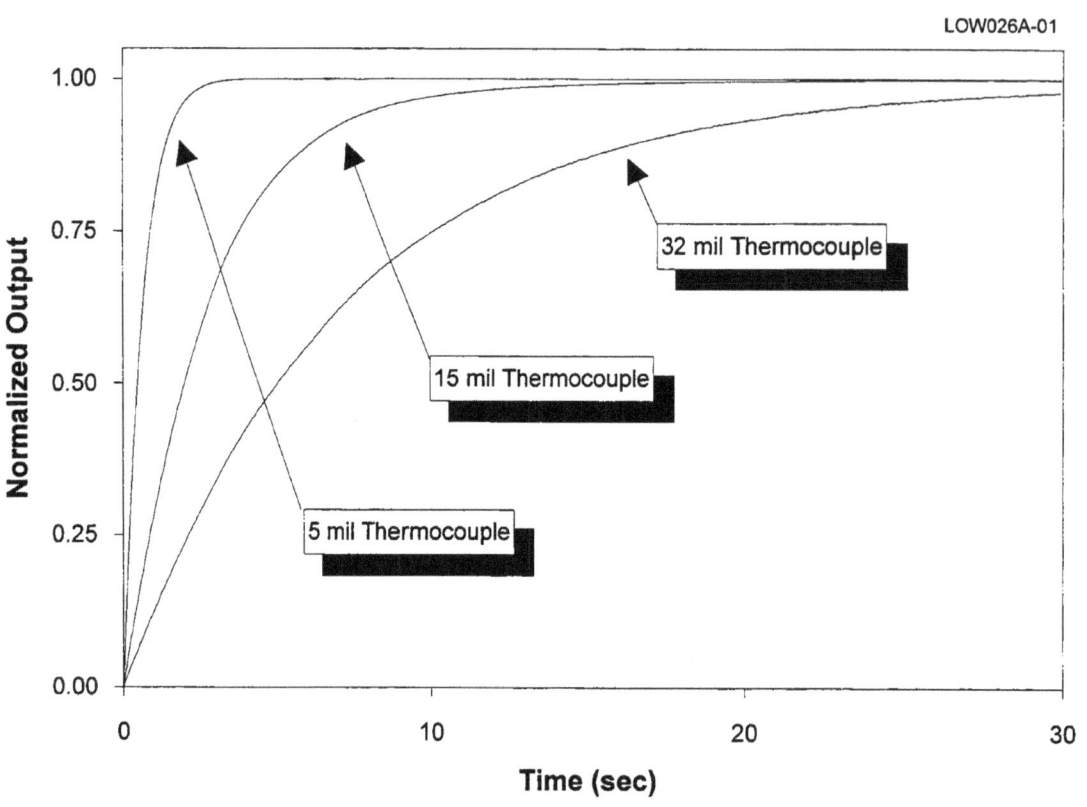

Figure 6.6 Comparison of LCSR Transients For Different Size Thermocouples

thermocouple wires, Peltier heating/cooling is concentrated at the junctions of dissimilar metals in the thermocouple circuit. Consequently, if a DC current is used, all the junctions in the circuit that have accumulated Peltier heating/cooling during the LCSR test will produce temperature transients after the current is cut off. These transients are unrelated to the response of the thermocouple junction and may cause error in the LCSR results. Furthermore, the Peltier heating/cooling at the measuring junction will result in axial as opposed to radial heat transfer. This is detrimental to the LCSR transient data analysis which is based on the assumption of predominantly radial heat transfer. In Joule heating, the heat transfer from the junction is predominantly radial. This is because with Joule heating, in addition to the junction, the thermocouple wires will heat up during the LCSR test. When the current is cut off, very little of the Joule heat at the junction can conduct axially through the wires because the wires are approximately as hot as the junction itself. This forces most of the heat to dissipate radially. With Peltier heating/cooling, however, there is a temperature gradient between the junction and the thermocouple wires when the current is cut off. Therefore, with Peltier heating/cooling, the heat is allowed to dissipate axially as well as radially.

To avoid the Peltier effect, AC currents are often used. The higher the frequency of the AC current, the lower is the Peltier effect. This is because the heating or cooling that is produced when the current travels in a given direction is canceled by the heating or cooling that is produced when the direction of the current is reversed. In order to minimize or avoid the Peltier effect, a 1000 Hz current source was used in the LCSR test equipment. Figure 6.7 illustrates the effects of Peltier

heating or cooling on the LCSR test transient for a thermocouple.[14]

6.3   LCSR Theory and Procedure

Background

The LCSR test is based on the principle that the output of a thermocouple to a step change in temperature induced inside the thermocouple can be converted to give the equivalent response for a step change in temperature outside the thermocouple (Figure 6.8). This is possible because the transfer function that represents the response to an external step change in temperature is related to that for an internal step change in temperature as follows:

$$G_{Plunge} = \frac{1}{(s - p_1)(s - p_2) \ldots} \tag{6.1}$$

$$G_{LCSR} = \frac{1}{(s - p_1)(s - p_2) \ldots} [(s - z_1)(s - z_2) \ldots] \tag{6.2}$$

Where G_{Plunge} represents the response that will be obtained in a plunge test and G_{LCSR} represents the response that will be obtained in a LCSR test. It is clear that the plunge response is a subset of the LCSR response meaning that if the LCSR response is known, then $p_1, p_2, \ldots$ will be known and can be used to obtain G_{Plunge}. The derivations that follow are carried out to show how Equations 6.1 and 6.2 were identified.

Heat Transfer Analysis of a Thermocouple System

The derivation of the LCSR and plunge test transfer functions given as G_{LCSR} and G_{Plunge}

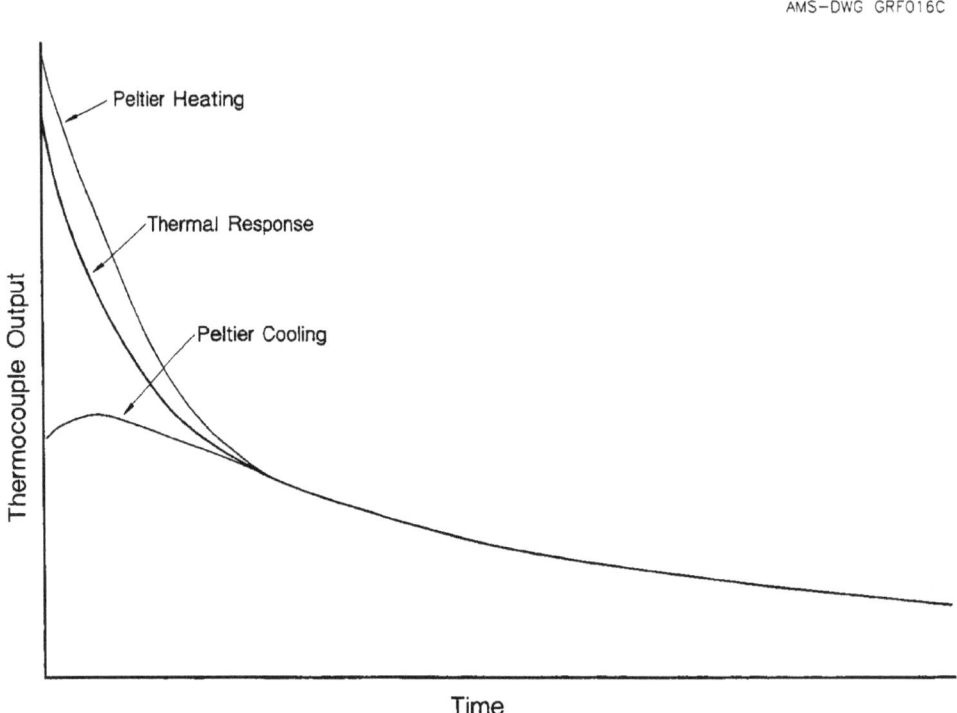

Figure 6.7 Peltier Effect On LCSR Test Transient

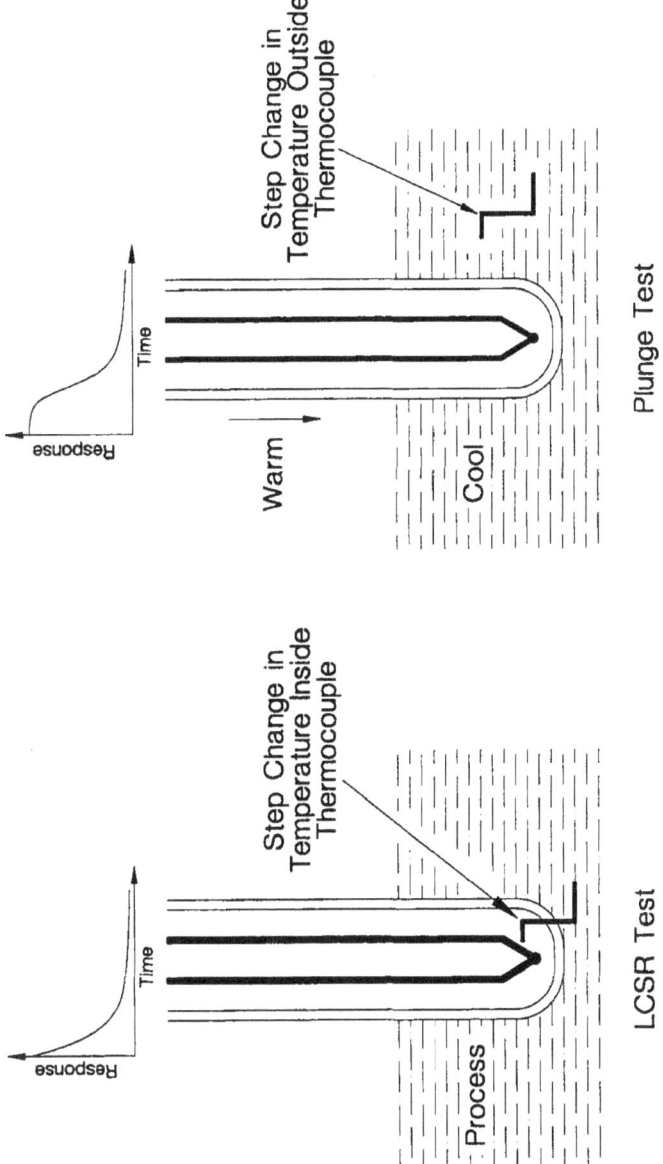

Figure 6.8 Comparison Of LCSR Method With Plunge Test

above are based on the assumption that the heat transfer between the thermocouple junction and the surrounding medium is one dimensional (radial). With this assumption, the heat transfer between the hot junction and the medium (fluid) surrounding the thermocouple may be represented by a lumped parameter network such as the one shown in Figure 6.9. For this network, the transient heat transfer equation for node i is written as:[15]

$$ mc \frac{dT_i}{dt} = \frac{1}{R_1} (T_{i-1} - T_i) - \frac{1}{R_2} (T_i - T_{i+1}) \qquad (6.3) $$

where m and c are the mass and specific heat capacity of material in the node, and R_1 and R_2 are the heat transfer resistances. Equation 6.3 may be rewritten as:

$$ \frac{dT_i}{dt} = a_{i,i-1} T_{i-1} - a_{i,i} T_i + a_{i,i+1} T_{i+1} \qquad (6.4) $$

where:

$$ a_{i,i-1} = \frac{1}{mcR_1} $$

$$ a_{i,i} = \frac{1}{mc} \left(\frac{1}{R_1} + \frac{1}{R_2} \right) \qquad (6.5) $$

$$ a_{i,i+1} = \frac{1}{mcR_2} . $$

The nodal equations may be applied to a series of nodes, starting with the node closest to the center ($i = 1$) and ending with the node closest to the surface ($i = n$):

$$ \frac{dT_1}{dt} = -a_{11} T_1 + a_{12} T_2 $$

$$ \frac{dT_2}{dt} = a_{21}T_1 - a_{22}T_2 + a_{23}T_3 $$

$$ \frac{dT_3}{dt} = a_{32}T_2 - a_{33}T_3 + a_{34}T_4 \qquad (6.6) $$

$$ \vdots $$

$$ \frac{dT_n}{dt} = a_{n,n-1}T_{n-1} - a_{n,n}T_n + a_{nF} T_F $$

where

T_i = temperature of the ith node (measured relative to the initial fluid temperature).

T_F = change of fluid temperature from its initial value.

These equations may be written in matrix form:
where:

$$ \frac{d\bar{x}}{dt} = A\bar{x} + \bar{f} T_F \qquad (6.7) $$

$$ \bar{x} = \begin{bmatrix} T_1 \\ T_2 \\ T_3 \\ \cdot \\ \cdot \\ \cdot \\ \cdot \\ T_n \end{bmatrix} \quad A = \begin{bmatrix} -a_{11} & a_{12} & 0 & 0 & 0 & 0 \\ a_{21} & -a_{22} & a_{23} & 0 & 0 & 0 \\ 0 & a_{32} & -a_{33} & a_{34} & 0 & 0 \\ 0 & \cdot & \cdot & \cdot & \cdot & \cdot \\ 0 & \cdot & \cdot & \cdot & \cdot & \cdot \\ 0 & \cdot & \cdot & \cdot & \cdot & \cdot \\ 0 & \cdot & \cdot & \cdot & a_{n,n-1} & -a_{n,n} \end{bmatrix} \quad \bar{f} = \begin{bmatrix} 0 \\ 0 \\ 0 \\ \cdot \\ \cdot \\ \cdot \\ a_{nF} \end{bmatrix} $$

Laplace transformation gives:

$$ [sI - A] \bar{x}(s) = \bar{f} T_F(s) + \bar{x}(t = 0). \qquad (6.8) $$

The solution for the temperature at the central node, $x_1 (s)$, is found by Cramer's rule:

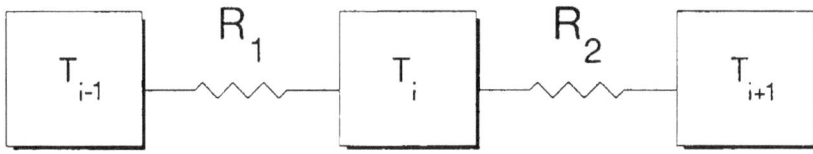

Figure 6.9 Lumped Parameter Representation For LCSR Analysis

$$T_1(s) = \frac{B(s)}{|sI-A|} \qquad (6.9)$$

where:

$$B(s) = \begin{bmatrix} T_1(0) & a_{12} & 0 & . & . & 0 \\ T_2(0) & (s+a_{22}) & -a_{23} & 0 & . & \\ T_3(0) & -a_{32} & (s+a_{33}) & -a_{33} & . & \\ . & & & & . & \\ . & & & & . & 0 \\ . & & & & . & 0 \\ [T_n(0)+a_n T_F(s)] & 0 & 0 & 0 & ... & -a_{n,n-1}(s+a_{n,n}) \end{bmatrix}$$

This Laplace transform is general for one-dimensional problems and its accuracy depends on the number of nodes used. Equation 6.8 is solved below for two different initial conditions, one initial condition to correspond to the LCSR test and the other to correspond to the plunge test. In the LCSR test, the temperature in the center node (hot junction of thermocouple) is not ambient at time t = 0, while for the plunge test, the temperature at the center node is ambient at t = 0.

LCSR Equation

For the LCSR test, $\bar{x}(t = 0)$ is the initial temperature distribution, and it is a vector with all entries nonzero, meaning that the first column of $B(s)$ in matrix 6.9 has all nonzero entries.

Evaluation of the determinants, $B(s)$ and $|sI-A|$, in Eq. 6.9 gives:

$$G(s) = \frac{T_1(s)}{T_F(s)}$$
$$= K \frac{(s - z_1)(s - z_2) \ldots (s - z_{n-1})}{(s - p_1)(s - p_2) \ldots (s - p_n)} \qquad (6.10)$$

where each z_i is a zero (a number that causes $T_1(s)$ to equal zero), and p_i is a pole (a number that causes $T_1(s)$ to equal infinity) and K is a constant gain factor that can be set

equal to unity to simplify the equation. The response $T_1(t)$ for a step change is obtained using the residue theorem (assuming all distant poles):

$$T_1(t) = \frac{(-z_1)(-z_2) \ldots (-z_{n-1})}{(-p_1)(-p_2) \ldots (-p_n)}$$
$$+ \frac{(p_1 - z_1)(p_1 - z_2) \ldots (p_1 - z_{n-1})}{(p_1 - p_2)(p_1 - p_3) \ldots (p_1 - p_n)} e^{p_1 t} \qquad (6.11)$$
$$+ \frac{(p_2 - z_1)(p_2 - z_2) \ldots (p_2 - z_{n-1})}{(p_2 - p_1)(p_2 - p_3) \ldots (p_2 - p_n)} e^{p_2 t} + \ldots$$

This may be rewritten as

$$T(t) = A_0 + A_1 e^{p_1 t} + A_2 e^{p_2 t} + \ldots \qquad (6.12)$$
$$A_0, A_1, A_2, \ldots = f(p_1, p_2, \ldots z_1, z_2, \ldots).$$

Equation 6.10 is referred to as the LCSR transfer function (G_{LCSR}) and Equation 6.12 is referred to as the equation for the LCSR transient. If the data for a LCSR test is mathematically fit to Equation 6.12, the values of $p_1, p_2, \ldots$ can be identified and used to construct the plunge test transient.

Plunge Test Equation

For a step perturbation of fluid temperature, $T_F(s)$ is nonzero, but $\bar{x}(t = 0)$ has all zero entries because the initial temperature distribution is flat and equal to the initial fluid temperature. In this case, the first column of $B(s)$ contains all zeros, except for the last entry.

In this case, $B(s)$ from matrix 6.9 may be written as:

$$B(s) = \begin{bmatrix} 0 & a_{12} & 0 & . & . & \\ 0 & (s+a_{22}) & -a_{23} & 0 & . & . \\ 0 & -a_{32} & (s+a_{33}) & -a_{34} & . & . \\ . & & & & . & \\ . & & & & . & \\ . & & & & . & 0 \\ a_{nF} T_F(s) & 0 & 0 & 0 & . & -a_{n,n-1}(s+a_{n,n}) \end{bmatrix}$$

Using the Laplace expansion method for evaluation of the determinants, we obtain:

$$B(s) = a_{nF} T_F(s)(-1)^{n+1} \begin{vmatrix} -a_{12} & 0 & 0 & 0 & \cdots \\ (s+a_{22}) & -a_{23} & 0 & 0 & \cdots \\ -a_{32} & (s+a_{33}) & -a_{34} & 0 & \cdots \\ 0 & -a_{43} & (s+a_{44}) & -a_{45} & \cdots \\ \cdot & \cdot & \cdot & & \cdots \\ \cdot & \cdot & \cdot & & \cdots \\ \cdot & \cdot & \cdot & & \cdots \end{vmatrix}$$

This is a lower diagonal matrix, and its determinant is the product of the diagonals:

$$B(s) = a_{nF} T_F(s)(-1)^{n+1}(a_{12}\, a_{23}\, a_{34}\, \cdots) \quad (6.13)$$

Therefore:

$$T_1(s) = \frac{a_{nF} T_F(s)(-1)^{n+1}}{(s-p_1)(s-p_2)\,\cdots\,(s-p_n)} \quad (6.14)$$

and the transfer function $\dfrac{T_1(s)}{T_F(s)}$ is:

$$G(s) = \frac{K}{(s-p_1)(s-p_2)\,\cdots\,(s-p_2)} \quad (6.15)$$

where K is a constant that can be set equal to unity to simplify the equation. By using the residue theorem, we obtain the following expression for the fluid temperature step change (Laplace transform of a unit step, i.e., $T_F(s) = \dfrac{1}{s}$):

$$\begin{aligned}
T_1(t) = & \frac{1}{(-p_1)(-p_2)\,\cdots\,(-p_n)} \\
& + \frac{1}{p_1(p_1-p_2)(p_1-p_3)\,\cdots\,(p_1-p_n)} e^{p_1 t} \\
& + \frac{1}{p_2(p_2-p_1)(p_2-p_3)\,\cdots\,(p_2-p_3)} e^{p_2 t} + \cdots
\end{aligned}$$

$$(6.16)$$

This equation may be written as:

$$T_1(t) = B_0 + B_1 e^{p_1 t} + B_2 e^{p_2 t} + \cdots \quad (6.17)$$

$$B_0,\, B_1,\, B_2\, \cdots = f(p_1, p_2, \cdots)$$

The following observations can be made about the fluid temperature step change (plunge) case:

1. The exponential terms $(p_1, p_2, \cdots)$ are the same as those of the LCSR result. This is expected since the exponents depend only on the heat transfer resistances and heat capacities, and these are the same in both cases.

2. The coefficients that multiply the exponentials are determined by the values of the poles but not of the zeros. Therefore, a knowledge of the poles alone is sufficient to determine these coefficients and the exponentials.

LCSR Transformation Procedure

The results of the derivations carried out above are used with the following procedure to convert the LCSR transient to give the equivalent plunge test transient:

1. Perform a LCSR test and sample the data with a computer. Normally, the LCSR transient starts at a high output value when the LCSR test begins and decreases as the thermocouple cools to the ambient temperature. However, it is customary to invert the LCSR transient and show it from low to high.

2. Fit the LCSR data to the following equation and identify the p_i's. The A_i's do not have to be identified.

$$T(t) = A_0 + A_1 e^{p_1 t} + A_2 e^{p_2 t} + \ldots \quad (6.18)$$

3. Use the p_i's identified above in Equation 6.18 to construct the temperature response that would have occurred if a fluid temperature step had been imposed.

4. Use the transient identified in step 3 to obtain the time constant of the thermocouple by determining the time that it takes for the transient to reach 63.2 percent of its final steady state value. Another approach, which is more often used to obtain the time constant, involves substituting the p_1, p_2, . . . (or τ_1, τ_2, τ_3, . . .) in Equation 6.19 to obtain the time constant directly.

$$\tau = \tau_1 [1 - \ln(1 - \frac{\tau_2}{\tau_1}) - \ln(1 - \frac{\tau_3}{\tau_1}) \ldots] \quad (6.19)$$

Two Dimensional Heat Transfer

The approach used above can be followed to analyze the thermocouple heat transfer based on a two dimensional model.[15] The reader may consult Reference 15 for a derivation of the two dimensional equation. The key results of the two dimensional analysis is that, unlike the one dimensional case, the step response results have zeros in the transfer function as well as poles. That is, the poles identified by the LCSR test are not all that is needed to construct the plunge test results. However, experience with typical thermocouples in typical installations has shown that the errors due to a minor departure from one dimensional assumptions are often not significant.

LCSR Test Procedure

Appendix A includes a general LCSR procedure for testing of thermocouples. This procedure was written for field measurements during a project that was conducted for the U.S. Air Force.

Raw Data

Numerous plunge and LCSR tests were performed during the Phase I project as described later in this report. Samples of the raw LCSR and plunge tests conducted in Phase I are included in Appendix B.

7. DEVELOPMENT OF THERMOCOUPLE FLOW SENSORS

Thermocouple flow sensors were developed during this project using Type K thermocouples. The details are presented in this section.

7.1 Thermocouple Selection

Thermocouples are referred to as sheathed, unsheathed, and grounded junction depending on how the thermocouple is constructed. Figure 7.1 illustrates the three styles. Exposed junction thermocouples are also referred to as unsheathed thermocouples. The response times of unsheathed thermocouples are the most sensitive to flow.

The thermocouple junction can be made in many ways. Figure 7.2 shows four examples of exposed junction configurations. Figure 7.3 shows typical LCSR transients for a sheathed and an unsheathed thermocouple of the same wire size (10 mil). Note that the unsheathed thermocouple has a much faster dynamic response than the sheathed thermocouple.

A number of unsheathed thermocouples were selected for the new air flow sensor development in this project. Thermocouple specifications such as wire size, junction surface area, loop resistance, thermocouple type, and availability were the factors used to make the selection. The thermocouples selected for this project had the following wire diameters: 5, 10, 15, and 20 mils. These thermocouples were obtained from Omega Engineering Company of Stamford, Connecticut (Table 7.1).

7.2 Thermocouple Test Instrumentation

AMS has performed extensive response time testing of thermocouples in previous research and development projects. These previous projects have lead to the development of a test instrument called the ETC-1 which provides an alternating current (AC) to heat thermocouples for in-situ response time measurement using the LCSR test method. This AC current can be controlled, by the ETC-1, to simulate a step change in temperature. In addition, this instrument provides a method to control the time that the heating current is applied to the thermocouple. An external data acquisition system is used with the ETC-1 to sample the thermocouple's output after the heating current is stopped. This output is stored on a computer disk and subsequently analyzed using AMS proprietary software to give the response time of the thermocouple. A block diagram of the ETC-1 and the data acquisition system is shown in Figure 7.4.

7.3 Construction and Calibration of the Test Loop

The development of the new air flow sensor required that the sensor's indications be verified against a known standard. To provide a constant air velocity and to prevent outside air disturbances from affecting the test conditions, a closed air flow test loop was constructed and calibrated with a Laser Doppler Velocimeter (LDV). A photograph of the test loop is shown in Figure 7.5.

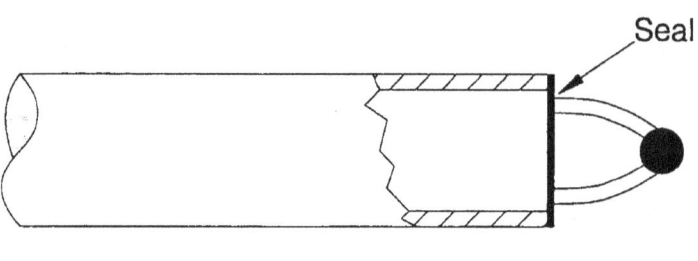

Exposed Junction

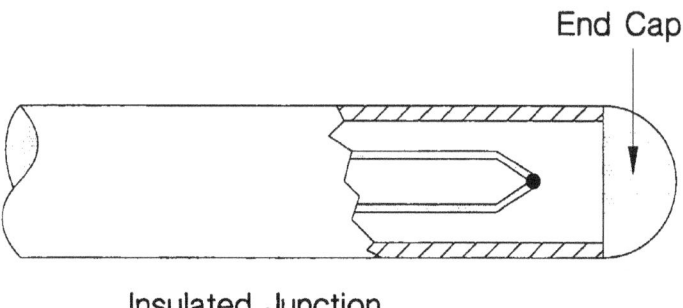

Insulated Junction

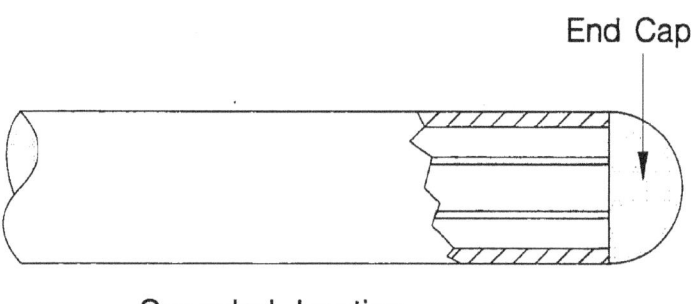

Grounded Junction

Figure 7.1 Typical Configurations of Measuring Junction
of Sheathed Thermocouples

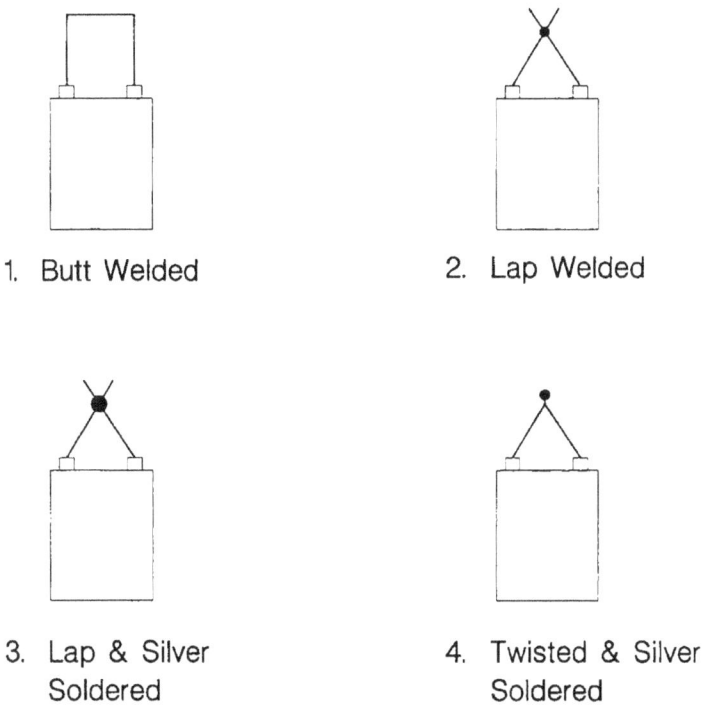

1. Butt Welded

2. Lap Welded

3. Lap & Silver
Soldered

4. Twisted & Silver
Soldered

Figure 7.2 Methods of Fabricating Thermocouple Junctions

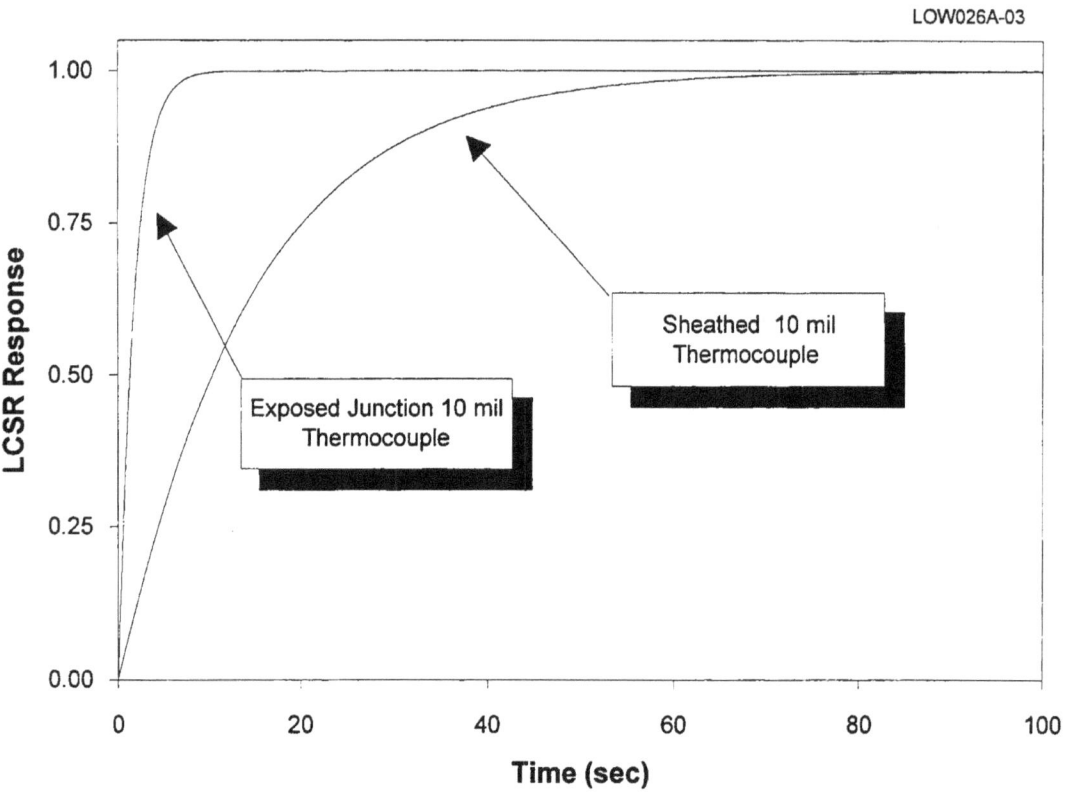

Figure 7.3   Typical LCSR Transients for a Sheathed
and Unsheathed Thermocouple

TABLE 7.1

Listing of Thermocouples Used in This Project

Item	Wire Diameter	Tag No.	Junction Surface Area (mils^2)	Loop Resistance (ohms)
1		N05A	125.0	24.6
2	5 mil	N05B	110.0	23.1
3		N05C	119.0	23.7
4		N05D	136.3	22.2
5		N10A	387.0	8.7
6	10 mil	N10B	446.5	7.7
7		N10C	360.0	8.6
8		N10D	484.0	9.5
9		N15A	2100.0	4.9
10	15 mil	N15B	1218.0	4.9
11		N15C	1470.0	5.0
12		N15D	1140.0	4.7
13		N20A	6160.0	1.5
14	20 mil	N20B	4721.4	1.3
15		N20C	6480.0	1.3
16		N20D	6160.0	1.6

Note: mil = 1/1000 inch

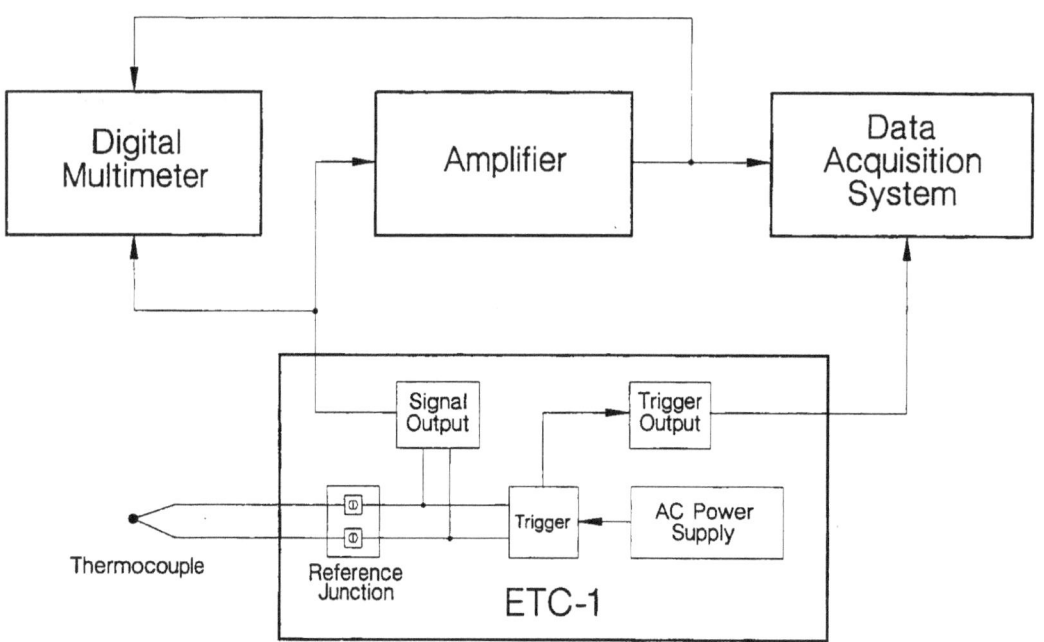

Figure 7.4 Block Diagram of the ETC-1 and the Data Acquisition System

Figure 7.5 Photograph of an Air Flow Test Loop

The air flow rates inside the test loop can be varied using a computer-controlled stepper motor. The stepper motor is connected by a drive belt to a fan mounted inside the test loop as shown in Figure 7.6. In addition, a drive pulley with three different turn-ratios is used to vary the air velocities produced by the fan motor. The speed of the fan is controlled by a computer that operates the stepper motor control electronics allowing the user to accurately reproduce selected air flow rates inside the loop.

The test loop provides a controlled environment for testing of thermocouples at velocities from near stagnant to 170 cm/s. In addition, the test loop provides a near uniform air flow profile across the inside diameter of the pipe by forcing the air around the entire length of the loop before coming into contact with the thermocouple under test.

The LDV that was used for the calibration of the test loop was made available for use by AMS in this project through a technology transfer agreement with ORNL. The test loop was transported to ORNL and calibrated against the LDV.

Air velocity measurements in the loop were made using the LDV. These measurements were then described as a function of the stepper motor RPM to calibrate the loop's air controller. The air velocity measurements were made by injecting an atomized glycerin solution, seed particles, into the test loop and measuring the velocity at the center of the loop piping using the LDV's laser beams. A photograph of the setup for these measurements is shown in Figure 7.7.

Measurements were performed at different motor RPM settings for each of the three drive pulleys. The results of these measurements are shown graphically in Figure 7.8. Using the information obtained from this calibration, the air flow controller can be used to accurately reproduce the air velocities measured by the LDV. In addition, this information will allow future laboratory flow measurements to be referenced to the LDV.

Figure 7.6   Detailed View of Stepper Motor and Fan
                Assembly for the Air Flow Loop

Figure 7.7 Photograph of Test Loop During Calibration with an LDV

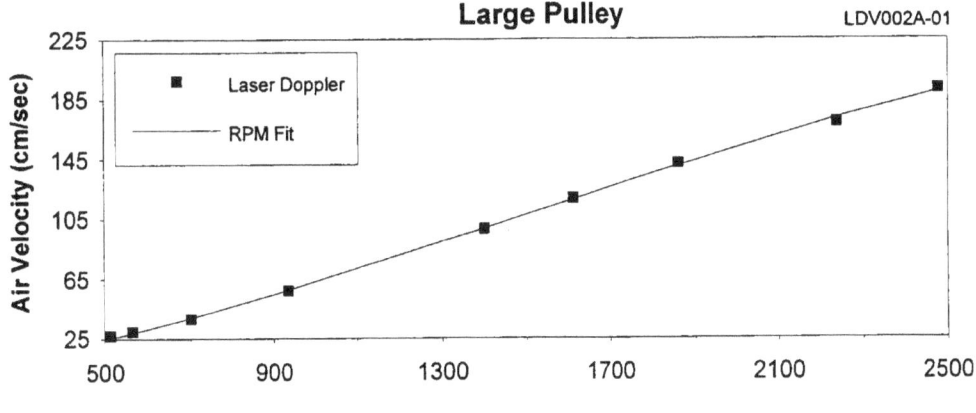

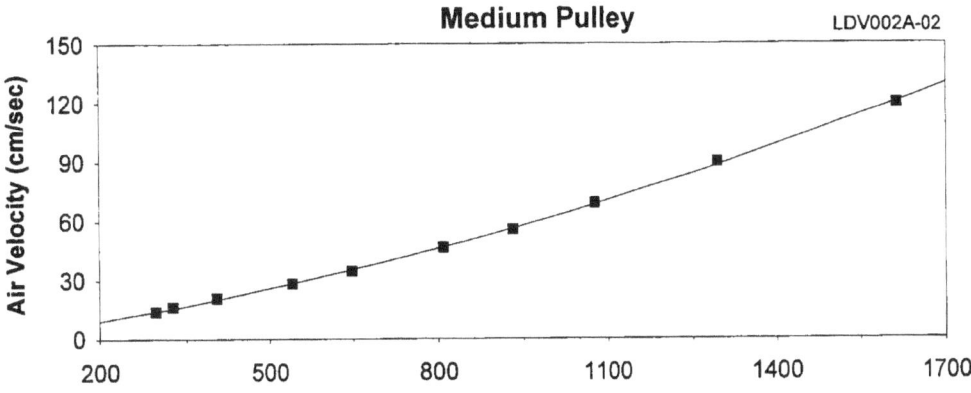

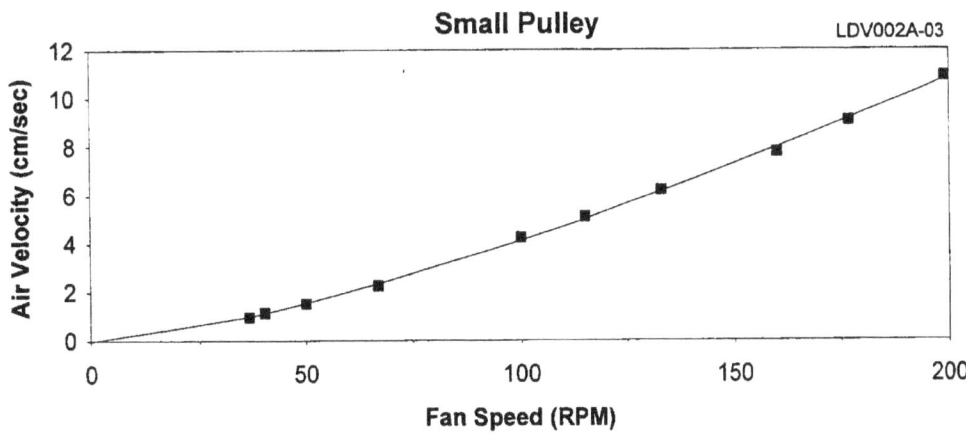

Figure 7.8 Results of Calibration of Test Loop Using the LDV

8. LABORATORY TEST RESULTS

Laboratory tests were performed as described in this section to demonstrate the feasibility of thermocouple flow sensors for measurement of ambient air flow rates. First, laboratory validation tests were performed to demonstrate how well the LCSR method provides the response time of thermocouples. This information is used in section 9 to discuss the accuracy of thermocouple flow sensors. The LCSR validation tests described below are followed by the laboratory results from the project showing the response time-versus-flow correlation data generated for several thermocouples.

8.1 LCSR Validation Tests

The validity and accuracy of the LCSR test method was established by performing response time measurements on thermocouples, both by LCSR and plunge tests, under identical test conditions.

Response time-versus-flow correlations were developed for Type K thermocouples of various sizes. The response time data were acquired using the standard plunge and LCSR test methods. A typical plunge and LCSR transient is shown in Figure 8.1. The plunge and LCSR tests were performed at air velocities ranging from 46 to 170 cm/s. The resulting data were subsequently analyzed to determine the response time (τ) of the thermocouple at the given air velocity. The response time results from the LCSR testing were determined by using several independent analysis programs, each of which uses a different algorithm to determine the response time of the thermocouple. The results from these programs were then averaged and used

as the final response time of the thermocouple. A comparison of the plunge and LCSR response time test results are shown in Table 8.1 and Figure 8.2 for several thermocouples that were tested.

The results in Table 8.1 and Figure 8.2 indicate a good agreement between the response time results obtained using the two methods. That is, the LCSR test can reproduce the plunge test response time of the thermocouple.

8.2 Identification of Optimum LCSR Test Parameters

A number of test parameters may effect the results of an LCSR test. These parameters must be identified for the tests to yield accurate response time results. The LCSR test parameters that may effect the dynamic response of a thermocouple are heating time and heating current. In order to determine the optimum test parameters for the LCSR testing, the effects of heating time and heating current on a thermocouple's dynamic response were examined.

A series of plunge and LCSR tests were performed at an air velocity of 17 m/s. First, plunge tests were performed in order to establish the thermocouple's baseline response time. LCSR tests were then performed while holding the heating current constant and varying the time that the heating current was applied to the thermocouple. Another series of LCSR tests were performed in which the heating time was held constant and various heating currents were applied to the thermocouple. The results of these tests

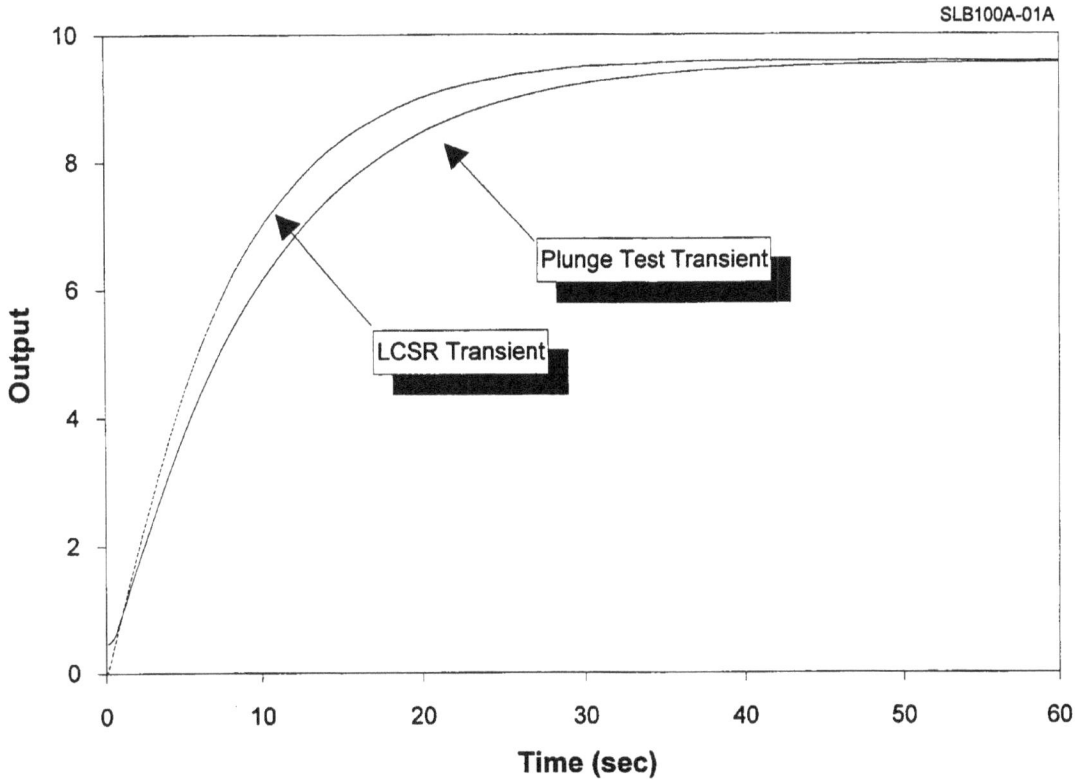

Figure 8.1 A Typical Plunge and LCSR Transient

TABLE 8.1

Comparison of Plunge and LCSR Response Time
Results for Various Thermocouple Sizes

Air Velocity (cm/s)	Response Time (sec)							
	5 mil		10 mil		15 mil		20 mil	
	Plunge	LCSR	Plunge	LCSR	Plunge	LCSR	Plunge	LCSR
46	1.1	0.9	2.5	2.3	4.5	5.0	6.4	7.4
60	1.0	0.9	2.2	2.3	3.9	4.6	5.7	6.8
76	0.9	0.8	2.0	2.1	3.6	4.3	5.2	6.2
96	0.9	0.7	1.7	1.8	3.4	3.8	4.9	5.4
110	0.9	0.7	1.6	1.7	3.3	3.5	4.6	5.0
125	0.8	0.7	1.6	1.7	3.1	3.5	4.4	5.1
146	0.8	0.6	1.5	1.7	2.8	3.4	4.3	4.7
170	0.7	0.6	1.4	1.5	2.7	3.2	3.9	4.4

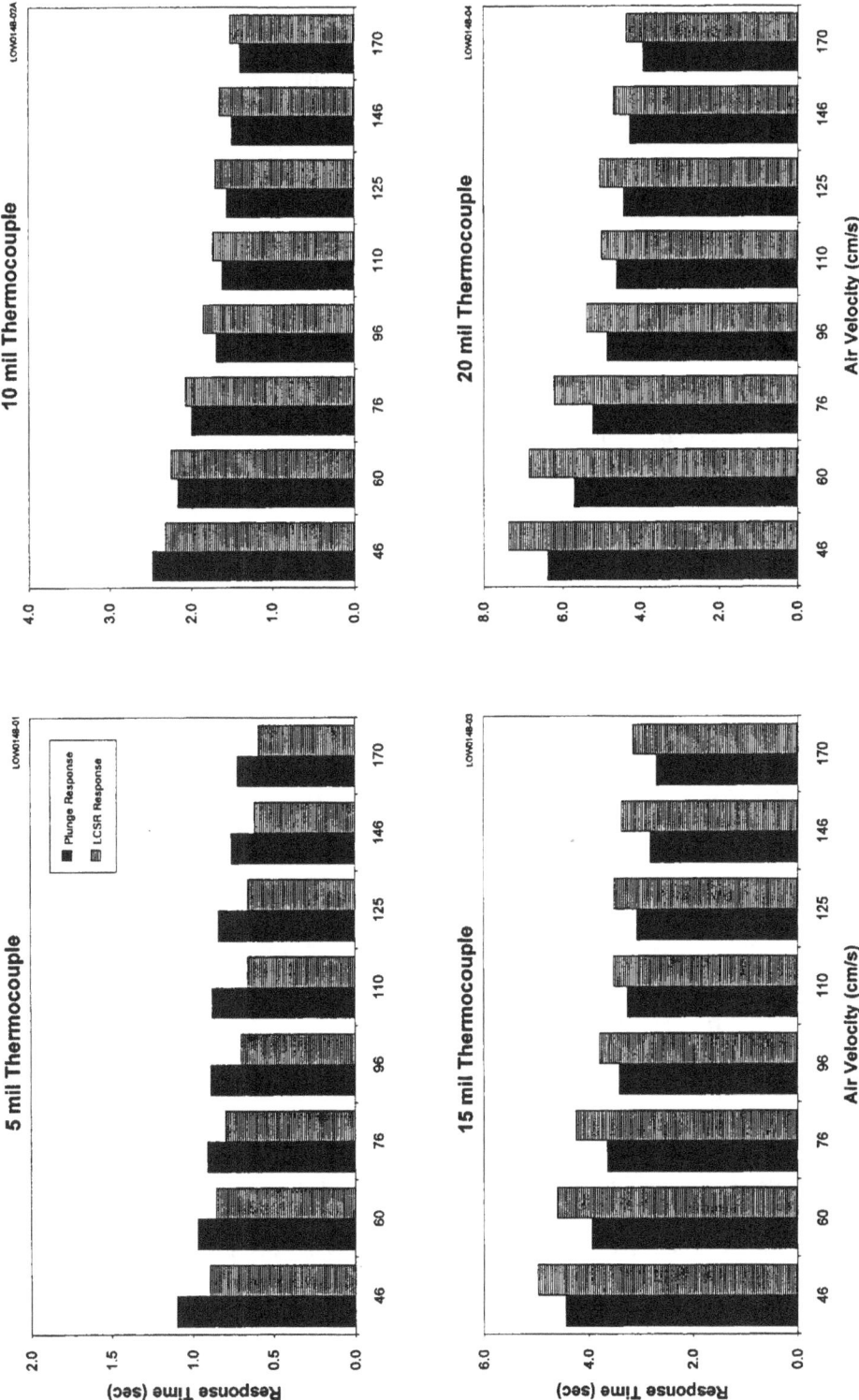

Figure 8.2 Comparison of Plunge and LCSR Test Results

shown in Table 8.2 for the variable heating time, and Table 8.3 for the variable heating current. These results show that the heating time applied to the thermocouple does not significantly affect the response time measurements. However, the magnitude of the heating current can influence the response time measurement. If the heating current applied to the thermocouple is too small, the measuring junction will not be thoroughly heated. This insufficient heating results in a slower response time than expected.

Figure 8.3 illustrates an LCSR transient where the measuring junction of the thermocouple is not completely heated followed by Figure 8.4 showing the effect on an LCSR transient when an excessive heating current is applied to the thermocouple. Note that the overshooting seen in the transient is due to extreme heating of the thermocouple wire.

8.3 Thermocouple Response Time Versus Air Flow Rate

Response time measurements were performed, using the LCSR method, in the test loop at air velocities from near stagnant to 170 cm/s. The test results are listed in Table 8.4 and shown in Figures 8.5 through 8.8. It is apparent from these results that the thermocouples are sensitive to small changes in air flow rates, especially at flow rates near zero. The 20 mil thermocouples were found to be the most sensitive to small changes in flow; however, they were also the slowest to respond to these changes. The 20 mil thermocouple took nearly 25 seconds to respond to low air velocity changes, while the 5 and 10 mil thermocouples responded to small changes in less than 7 seconds. This variation in response time is primarily due to the difference in

surface area of the thermocouple's measuring junction.

Response time variations were also noted among groups of thermocouples of the same size (Figure 8.9). These variations were found to be due to non-uniformity in the size of the measuring junction within each group. The variation of junction sizes are shown in Table 8.5.

8.4 Response Time Versus Air Flow Correlation

The Hilpert equation which is based on the heat transfer properties of fluids was initially used to convert the response time measurements to air velocities. A detailed description of these equations was given in Section 5. The equation has the following form:

$$\tau = C_1 + C_2 U^n \qquad (8.2)$$

where τ is the response time of the thermocouple and U is the air velocity. The equation constants C_1, C_2 and n were calculated by a least squares fitting method using the response time data. The constants for Equation 8.2 are given in Table 8.6 for each size thermocouple that was tested in this study.

Air velocity estimations were made using Equation 8.2. The results are listed in Table 8.7 and are shown graphically in Figure 8.10. This response versus flow correlation was reasonably accurate for air velocities from 10 to 170 cm/s, however, the correlation diverged below 5 cm/s. Table 8.8 lists the errors for each size of thermocouple that was tested and analyzed using Equation 8.2. It can be seen from this table that the errors for velocity estimations from 5 to 170 cm/s are relatively

TABLE 8.2

Constant Current and Varied Heating Time

Heating Time (sec)	Response Time (sec)
1.0	0.49
2.0	0.48
3.0	0.46
4.0	0.48
5.0	0.46

TABLE 8.3

Constant Heating Time and Varied Current

Heating Current (amps)	Response Time (sec)
0.50	0.49
0.75	0.50
1.00	0.46
1.25	0.43
1.50	0.43
1.75	0.35
2.00	0.35

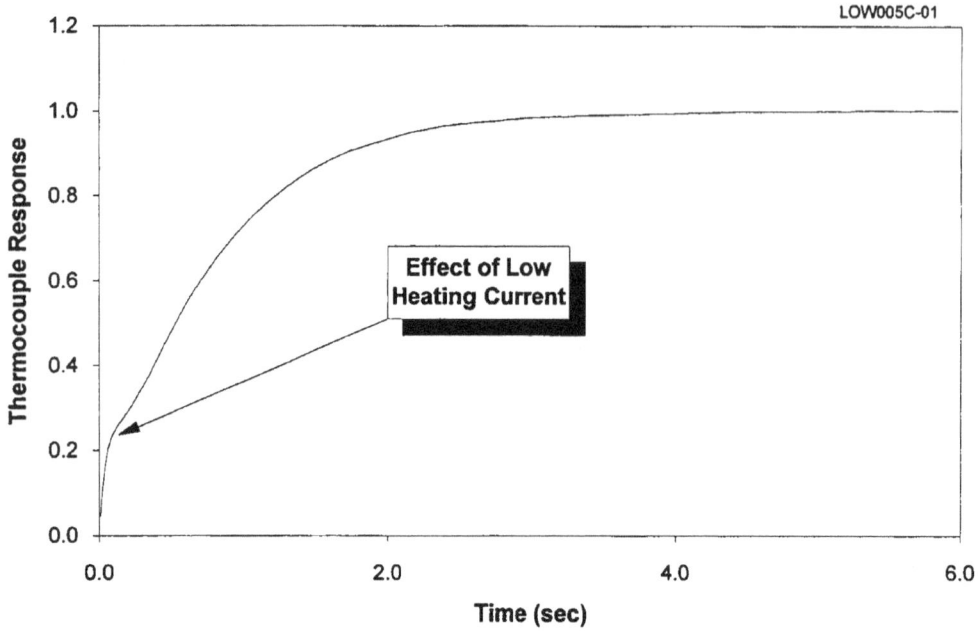

Figure 8.3 Illustration of Low Heating Current on an LCSR Transient

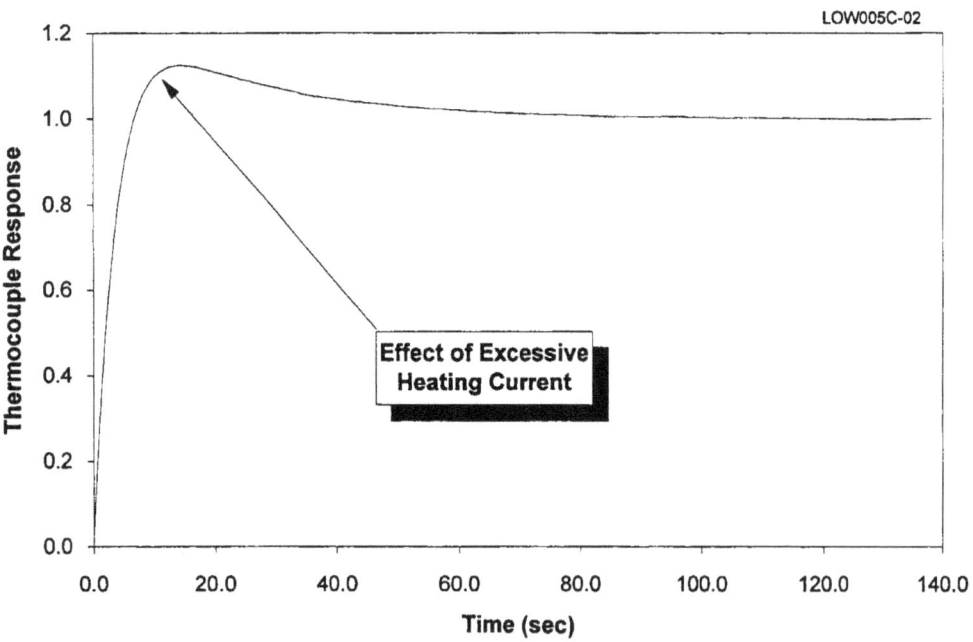

Figure 8.4 Illustration of Excessive Heating Current on an LCSR Transient

TABLE 8.4

Response Time of Thermocouples Versus Flow Rate

Actual Air Flow (cm/sec)	Response Time (sec)			
	5 mil	10 mil	15 mil	20 mil
0.97	2.18	6.50	14.64	25.05
1.12	2.17	6.41	14.39	24.71
1.52	2.10	6.16	13.65	23.24
2.24	1.95	5.64	12.50	20.75
4.24	1.71	5.01	10.81	18.27
5.10	1.64	4.83	10.47	17.43
6.21	1.54	4.59	10.08	16.69
7.75	1.41	4.37	9.60	15.61
9.05	1.34	4.24	9.35	14.91
10.86	1.26	4.07	9.02	14.14
14.09	1.17	3.67	8.19	12.28
16.11	1.13	3.54	8.05	11.54
20.66	1.05	3.21	6.94	10.51
27.04	0.99	2.84	6.48	9.44
28.20	0.96	2.83	5.98	9.17
29.86	0.95	2.81	5.80	8.95
34.37	0.91	2.59	5.45	8.15
38.16	0.87	2.55	5.26	7.85
46.50	0.82	2.32	4.98	7.36
55.30	0.77	2.27	4.63	6.86
57.37	0.76	2.25	4.62	6.85
68.70	0.70	2.07	4.26	6.23
89.50	0.63	1.89	3.88	5.57
98.47	0.60	1.85	3.79	5.39
118.57	0.57	1.74	3.61	5.05
119.20	0.56	1.71	3.51	5.03
141.98	0.52	1.65	3.36	4.69
169.30	0.48	1.53	3.14	4.35

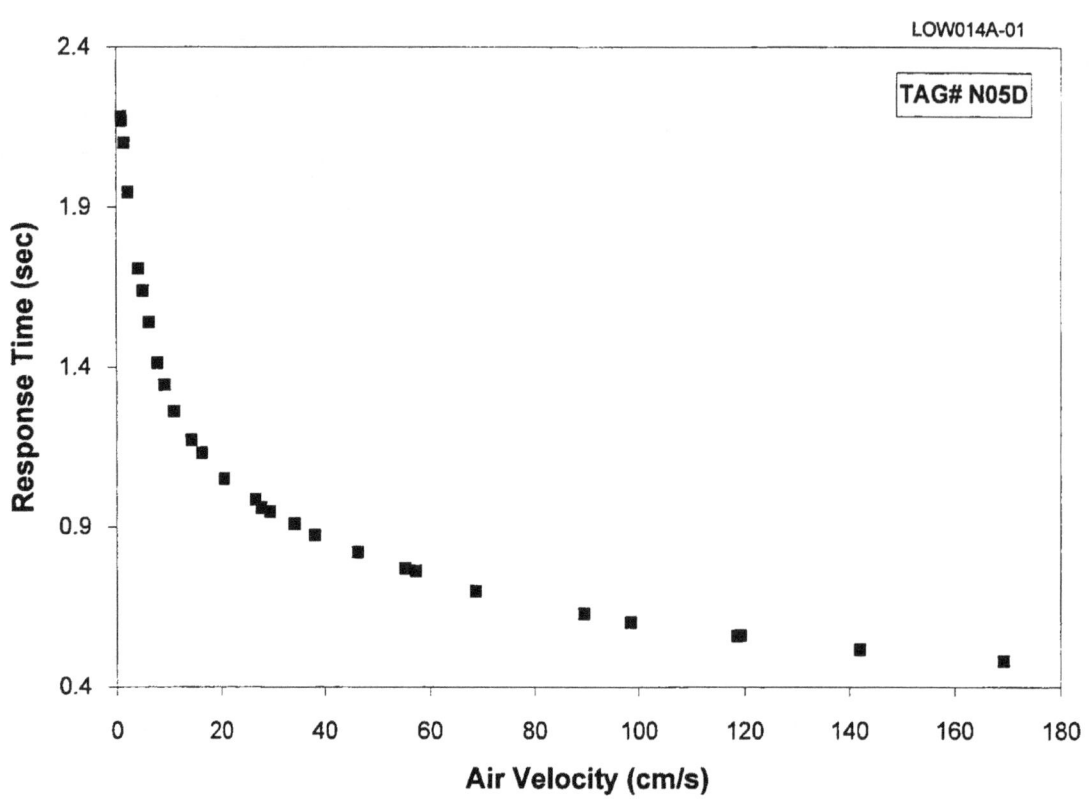

Figure 8.5 Response Time Versus Air Velocity for a 5 Mil Thermocouple

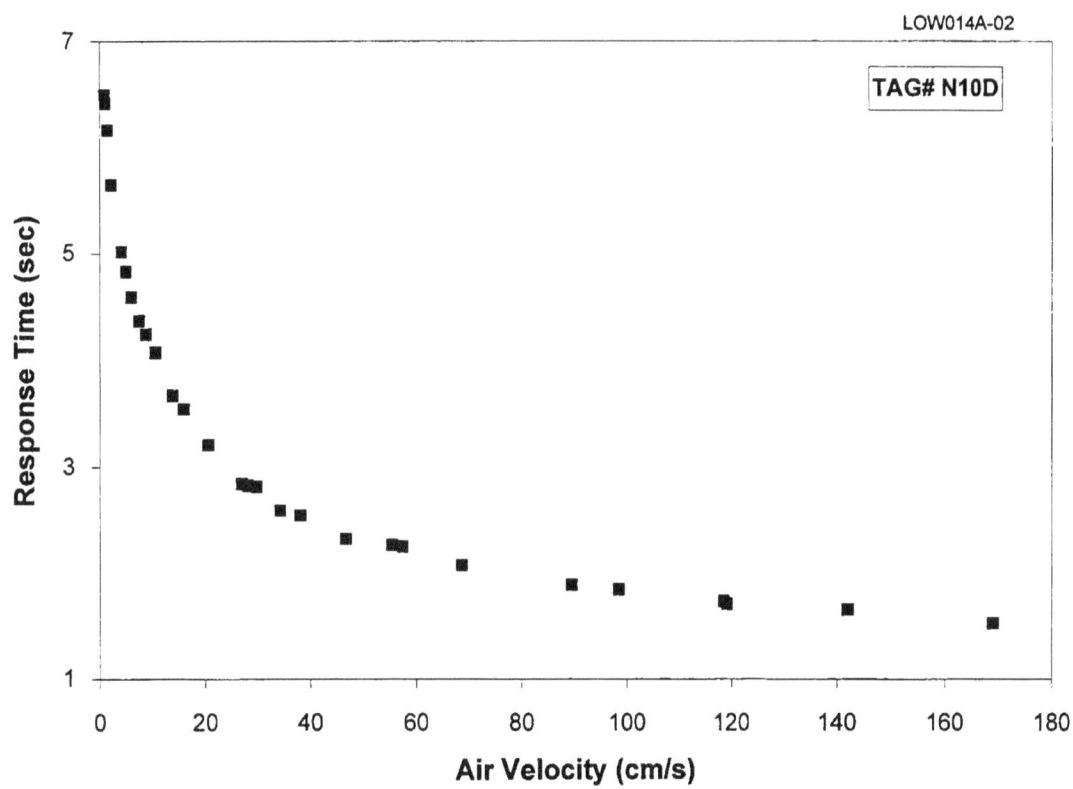

Figure 8.6 Response Time Versus Air Velocity for a 10 Mil Thermocouple

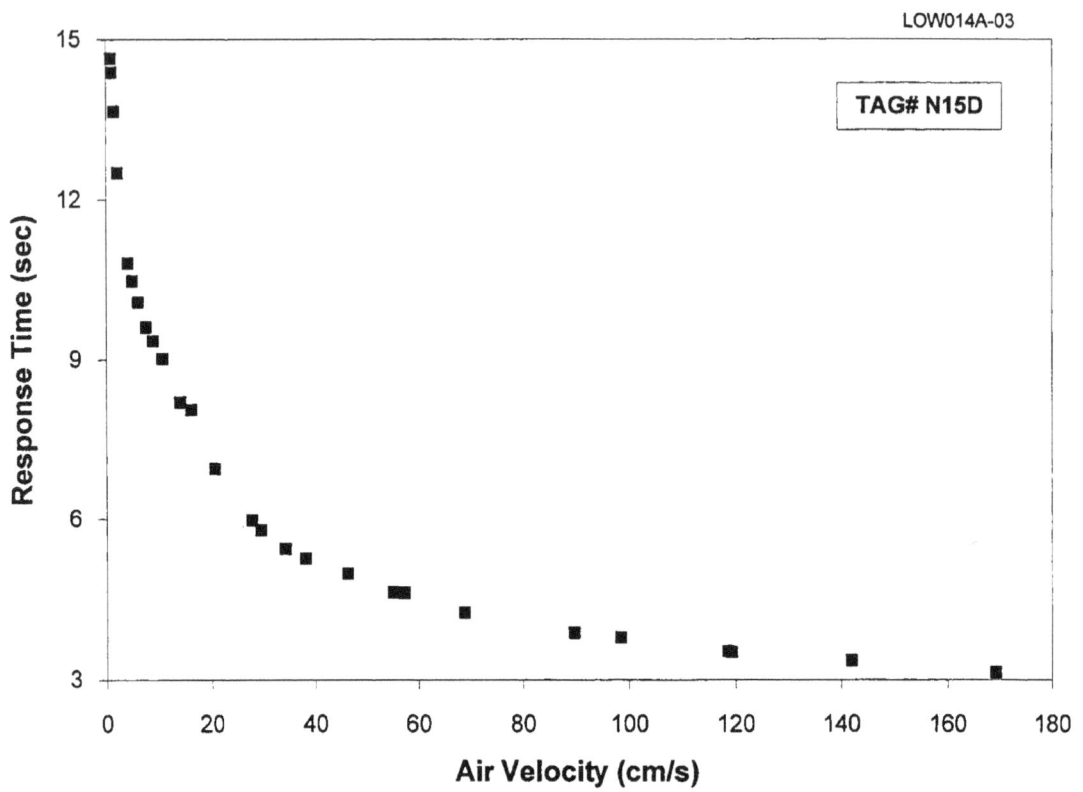

Figure 8.7 Response Time Versus Air Velocity for a 15 Mil Thermocouple

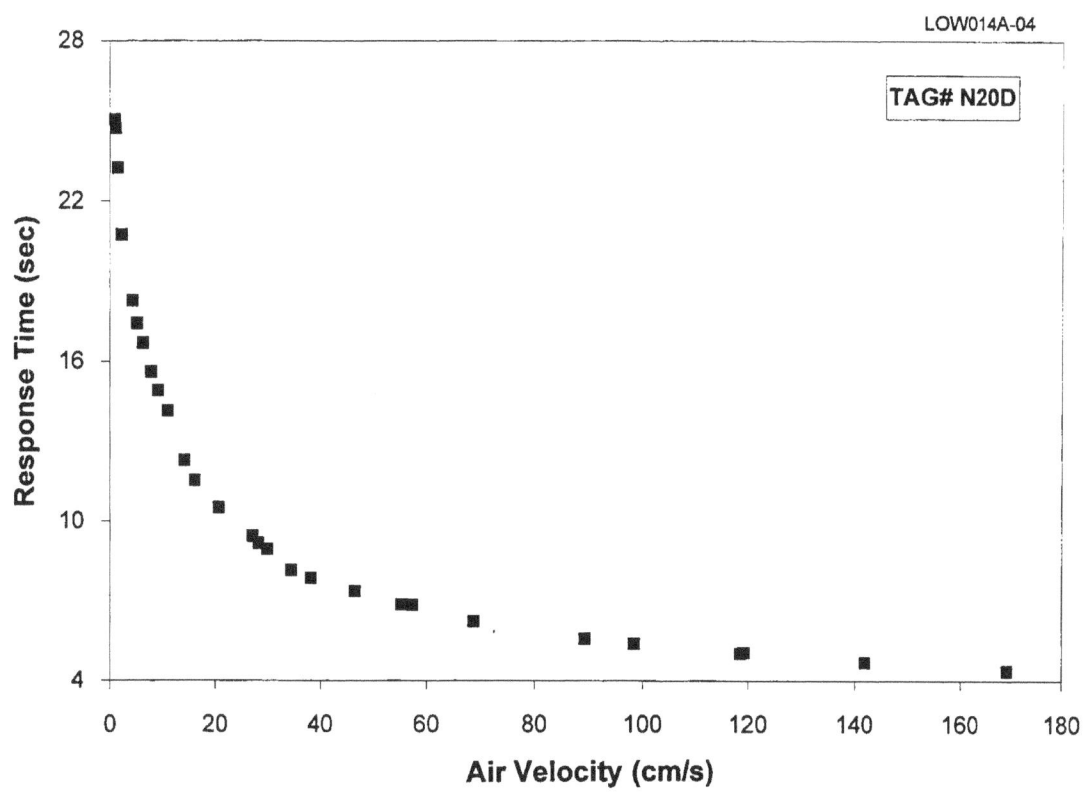

Figure 8.8 Response Time Versus Air Velocity for a 20 Mil Thermocouple

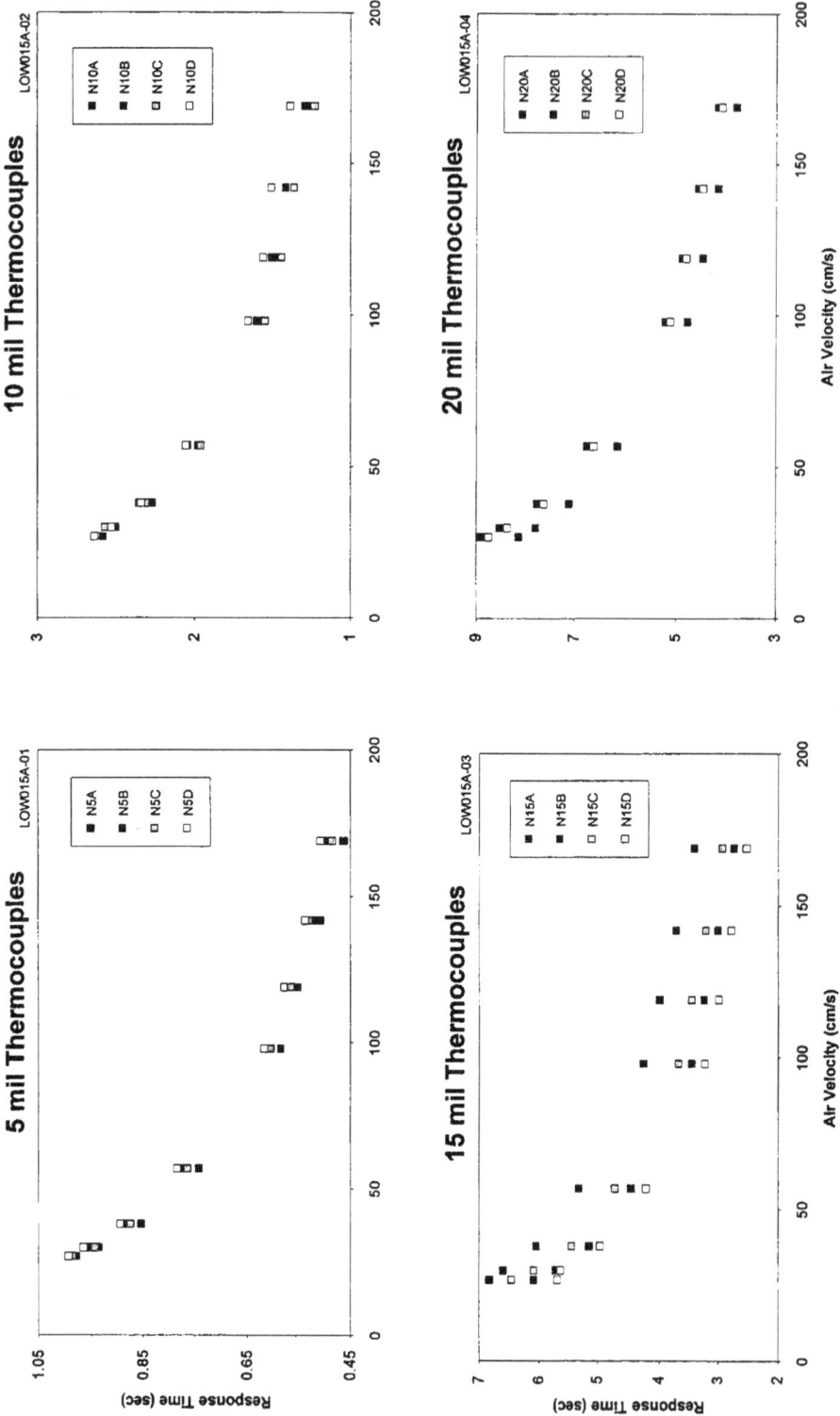

Figure 8.9 Response Time Variations Due to Junction Size Differences

TABLE 8.5

Differences in Thermocouple Junction Surface Area

T/C Tag No.	Dimensions (mils)		Surface Area (mils^2)	% Difference from Average
	X direction	Y direction		
N05A	12.5	10.0	125.0	2.0
N05B	11.0	10.0	110.0	-10.2
N05C	11.9	10.0	119.0	-2.9
N05D	10.9	12.5	136.3	11.2
Averages	**11.6**	**10.6**	**122.6**	
N10A	21.5	18.0	387.0	-7.7
N10B	23.5	19.0	446.5	6.5
N10C	20.0	18.0	360.0	-14.2
N10D	22.0	22.0	484.0	15.4
Averages	**21.8**	**19.3**	**419.4**	
N15A	50.0	42.0	2100.0	41.7
N15B	42.0	29.0	1218.0	-17.8
N15C	42.0	35.0	1470.0	-0.8
N15D	38.0	30.0	1140.0	-23.1
Averages	**43.0**	**34.0**	**1482.0**	
N20A	80.0	77.0	6160.0	4.8
N20B	61.0	77.4	4721.4	-19.7
N20C	81.0	80.0	6480.0	10.2
N20D	77.0	80.0	6160.0	4.8
Averages	**74.8**	**78.6**	**5880.4**	

TABLE 8.6

Coefficients for Response Versus Flow Correlation

$\tau=C_1+C_2U^n$				
TC Size	**U Range (cm/s)**		C_1	C_2	n
5 mil	0.0	25.0	0.1015	0.5655	0.330
	25.0	170.0	0.0044	0.5976	0.385
10 mil	0.0	14.0	1.0831	1.2483	0.330
	14.0	140.0	0.2425	1.5886	0.385
	140.0	170.0	0.0517	1.8843	0.466
15 mil	0.0	9.0	4.3866	2.2735	0.330
	9.0	90.0	-0.348	3.9648	0.385
	90.0	170.0	0.9927	2.7522	0.466
20 mil	0.0	4.0	7.2548	3.9232	0.330
	4.0	40.0	0.5962	5.4516	0.385
	40.0	170.0	0.9045	4.4714	0.466

TABLE 8.7

Velocity Measurements Using
Response Versus Flow Correlation

Actual Air Flow Rate (cm/sec)	Estimated Air Flow Rate (cm/s)			
	5 mil	10 mil	15 mil	20 mil
1.0	1.9	1.1	1.0	1.0
1.1	2.0	1.2	1.1	1.1
1.5	2.2	1.4	1.4	1.4
2.2	2.8	2.2	2.1	2.4
4.2	4.2	4.2	4.3	4.7
5.1	4.8	5.2	5.1	5.3
6.2	5.9	7.0	6.2	6.0
7.7	7.8	8.7	8.1	7.2
9.1	9.2	9.5	9.4	8.1
10.9	11.4	10.6	10.7	9.4
14.1	14.5	14.1	13.6	13.8
16.1	16.3	15.5	14.2	16.4
20.7	20.9	20.3	20.5	21.1
27.0	27.6	28.4	24.4	28.5
28.2	29.5	28.8	29.7	30.8
29.9	30.5	29.2	32.1	33.0
34.4	34.0	36.8	37.3	35.5
38.2	37.7	38.5	40.6	38.8
46.5	44.6	49.8	46.3	45.4
55.3	52.5	53.4	55.3	54.1
57.4	54.0	54.3	55.7	54.3
68.7	67.4	68.9	67.8	68.7
89.5	88.8	89.6	90.4	91.2
98.5	100.0	95.6	96.3	99.3
118.6	120.8	114.6	111.0	117.6
119.2	119.7	120.1	120.7	118.9
142.0	148.2	142.0	138.2	143.0
169.3	179.0	169.3	171.0	174.5

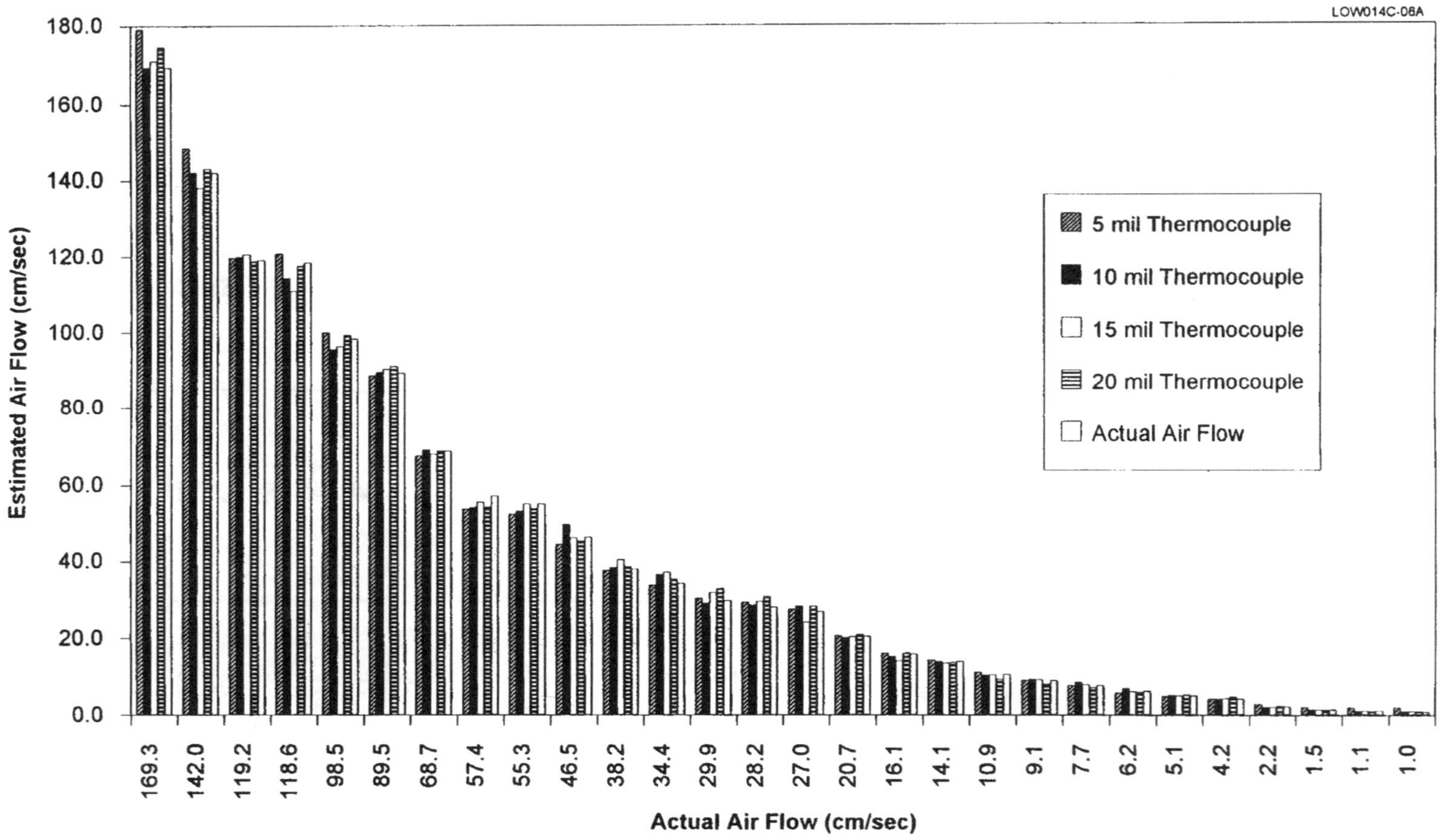

Figure 8.10 Comparison of Velocity Measurements Using Response Versus Flow Correlation

TABLE 8.8

Accuracy of Response Time Correlation

Actual Air Flow (cm/sec)	% Differences in Air Flow Estimation			
	5 mil	10 mil	15 mil	20 mil
1.0	98.3	13.5	7.2	5.5
1.1	74.7	4.2	0.2	3.5
1.5	43.8	6.3	6.6	6.6
2.2	23.9	1.7	5.7	5.5
4.2	0.4	1.6	1.3	11.0
5.1	5.2	1.2	0.8	4.8
6.2	4.8	12.6	0.2	3.1
7.7	0.6	12.9	4.1	7.1
9.1	1.7	4.6	3.5	10.0
10.9	4.8	2.7	1.2	13.4
14.1	3.2	0.2	3.2	2.0
16.1	1.3	4.0	11.6	1.7
20.7	1.0	1.5	0.6	2.3
27.0	1.9	4.9	9.9	5.4
28.2	4.8	2.0	5.4	9.4
29.9	2.2	2.3	7.3	10.7
34.4	1.2	6.9	8.6	3.3
38.2	1.1	0.8	6.3	1.7
46.5	4.1	7.1	0.4	2.3
55.3	5.0	3.5	0.1	2.2
57.4	5.9	5.4	3.0	5.3
68.7	1.9	0.3	1.3	0.0
89.5	0.8	0.2	1.0	1.9
98.5	1.6	2.9	2.2	0.9
118.6	1.9	3.4	6.4	0.8
119.2	0.4	0.8	1.2	0.3
142.0	4.4	0.0	2.7	0.7
169.3	5.7	0.0	1.0	3.1

low, however below 5 cm/s the errors increase significantly. The large errors below 5 cm/s are unacceptable for a new sensor. These errors are shown graphically in Figure 8.11.

One of the problems associated with Equation 8.2 is its complexity. The equation requires that several sets of constants be calculated for selected ranges of air velocities. Because of the complexity and large errors at low velocities, other mathematical solutions were sought. After evaluation, a logarithmic series was found to best describe the thermocouple's dynamic response for the flow rates of interest. This series is of the form

$$U = C_0 + C_1 ln(\tau) + C_2 ln(\tau)^2$$
$$+ C_3 ln(\tau)^3 + \dots + C_n ln(\tau)^n \qquad (8.3)$$

where τ is the response time of the thermocouple in seconds and U is the air velocity in cm/sec. This equation is complex, however it requires only one set of constants to cover the entire range of air velocities. The constants for the natural log equation are given in Table 8.9 for various size thermocouples used in this project. Air velocity estimations were calculated using Equation 8.3 and are given in Table 8.10 to one significant digit. The errors for air velocity using the series equation are shown in Table 8.11 and Figure 8.12. It is obvious that the series equation is capable of estimating air velocities with less than 10% error.

Once an accurate relationship is established for thermocouples it can be used for determining air velocities ranging from moderate flow to near stagnant conditions. The size of the thermocouple determines how fast the thermocouple responds to very low air velocities, however there is a trade-off between sensor sensitivity and response time.

8.5   Comparison of Hot Film Anemometer and Thermocouple Flow Sensor

The sensitivity of a hot film anemometer was compared with the new thermocouple flow sensor. This was done by measuring various air velocities using both sensors and comparing their outputs (Figures 8.13 and 8.14). Although the hot film anemometer has good sensitivity, it is not very useful for measurement of very low flows. This is because hot film anemometers create convective currents due to internal heating which can interfere with low air flow measurements.

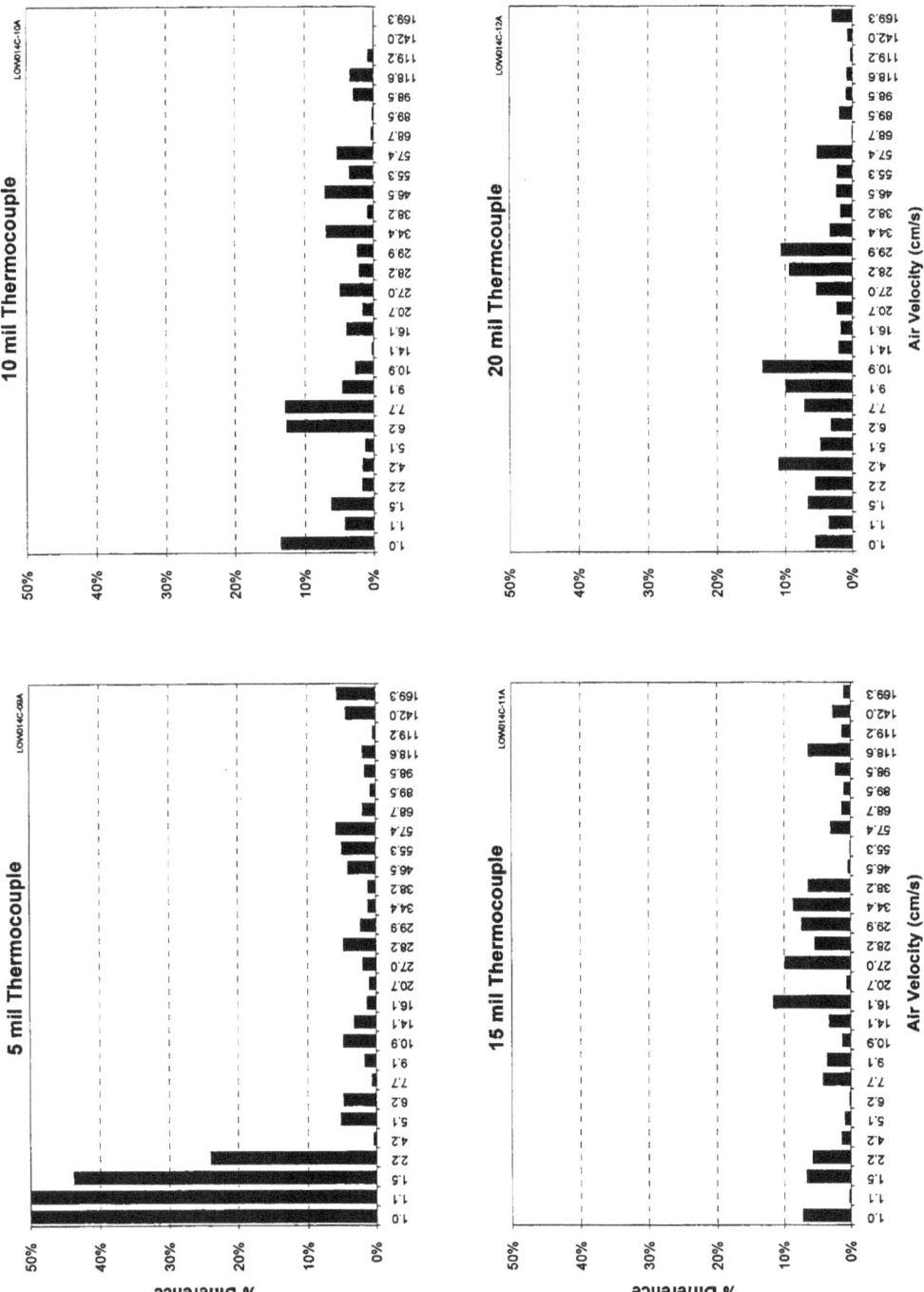

Figure 8.11 Errors Associated with Velocity Measurement Using the Heat Transfer Correlation

TABLE 8.9

Series Equation Constants

$$U = C_0 + C_1 ln(\tau) + C_2 ln(\tau)^2 + C_3 ln(\tau)^3 + ... + C_n ln(\tau)^n$$

	5 mil	10 mil	15 mil	20 mil
C_0	25.12	465.38	1168.26	1950.21
C_1	-88.83	-906.46	-685.64	-1474.84
C_2	111.80	403.16	-785.51	-921.56
C_3	67.25	324.80	283.02	1480.55
C_4	-63.50	-195.42	718.77	-675.75
C_5	-510.88	-220.82	-614.02	136.63
C_6	178.79	126.26	183.11	-10.52
C_7	818.77	168.18	-19.31	
C_8	-102.47	-204.68		
C_9	-503.53	83.52		
C_{10}		-12.35		

TABLE 8.10

Air Velocity Measurement Using Series Equation

Actual Air Flow Rate (cm/sec)	Estimated Air Flow Rate (cm/s)			
	5 mil	10 mil	15 mil	20 mil
1.0	1.0	1.0	1.0	0.9
1.1	1.1	1.1	1.1	1.1
1.5	1.6	1.5	1.5	1.5
2.2	2.2	2.2	2.1	2.2
4.2	4.3	4.1	4.5	4.4
5.1	5.1	5.1	5.3	5.2
6.2	6.2	6.5	6.4	6.1
7.7	7.8	8.0	8.0	7.6
9.1	9.0	9.0	9.0	8.8
10.9	10.9	10.4	10.4	10.2
14.1	14.1	14.4	14.3	14.6
16.1	16.0	15.8	15.0	16.9
20.7	21.1	20.5	21.5	20.7
27.0	26.4	28.0	24.9	26.2
28.2	28.9	28.4	29.4	27.9
29.9	30.2	28.7	31.5	29.5
34.4	34.4	36.1	36.1	36.3
38.2	38.9	37.8	39.1	39.5
46.5	46.7	49.3	44.8	45.9
55.3	55.1	53.0	54.3	54.5
57.4	56.6	54.0	54.7	54.7
68.7	69.7	69.5	69.0	69.1
89.5	89.6	91.6	90.7	91.5
98.5	99.8	98.0	96.9	99.5
118.6	118.6	118.0	111.9	117.3
119.2	117.6	123.7	121.6	118.5
142.0	142.7	136.3	138.7	141.2
169.3	169.2	170.5	169.0	169.6

TABLE 8.11

Accuracy of Measurements Using Series Equation

Actual Air Flow (cm/sec)	% Differences in Air Flow Estimation			
	5 mil	10 mil	15 mil	20 mil
1.0	0.4	1.8	0.9	2.5
1.1	1.9	2.0	2.0	2.2
1.5	3.0	1.1	0.6	1.1
2.2	0.9	2.0	7.9	1.9
4.2	0.4	2.2	5.1	2.6
5.1	1.0	0.9	3.6	2.0
6.2	0.0	4.3	3.5	1.9
7.7	1.1	3.7	3.7	1.9
9.1	0.7	0.5	0.6	3.3
10.9	0.6	4.1	4.4	6.1
14.1	0.1	1.9	1.3	3.7
16.1	0.5	1.9	6.9	4.8
20.7	1.9	0.9	4.1	0.4
27.0	2.4	3.5	7.9	3.1
28.2	2.6	0.6	4.4	1.1
29.9	1.0	3.7	5.3	1.4
34.4	0.2	5.0	5.1	5.6
38.2	2.1	1.0	2.5	3.5
46.5	0.4	6.0	3.7	1.3
55.3	0.4	4.2	1.8	1.5
57.4	1.4	5.9	4.6	4.6
68.7	1.4	1.1	0.4	0.5
89.5	0.1	2.4	1.4	2.2
98.5	1.4	0.5	1.6	1.1
118.6	0.0	0.5	5.6	1.1
119.2	1.4	3.8	2.0	0.6
142.0	0.5	4.0	2.3	0.6
169.3	0.1	0.7	0.2	0.2

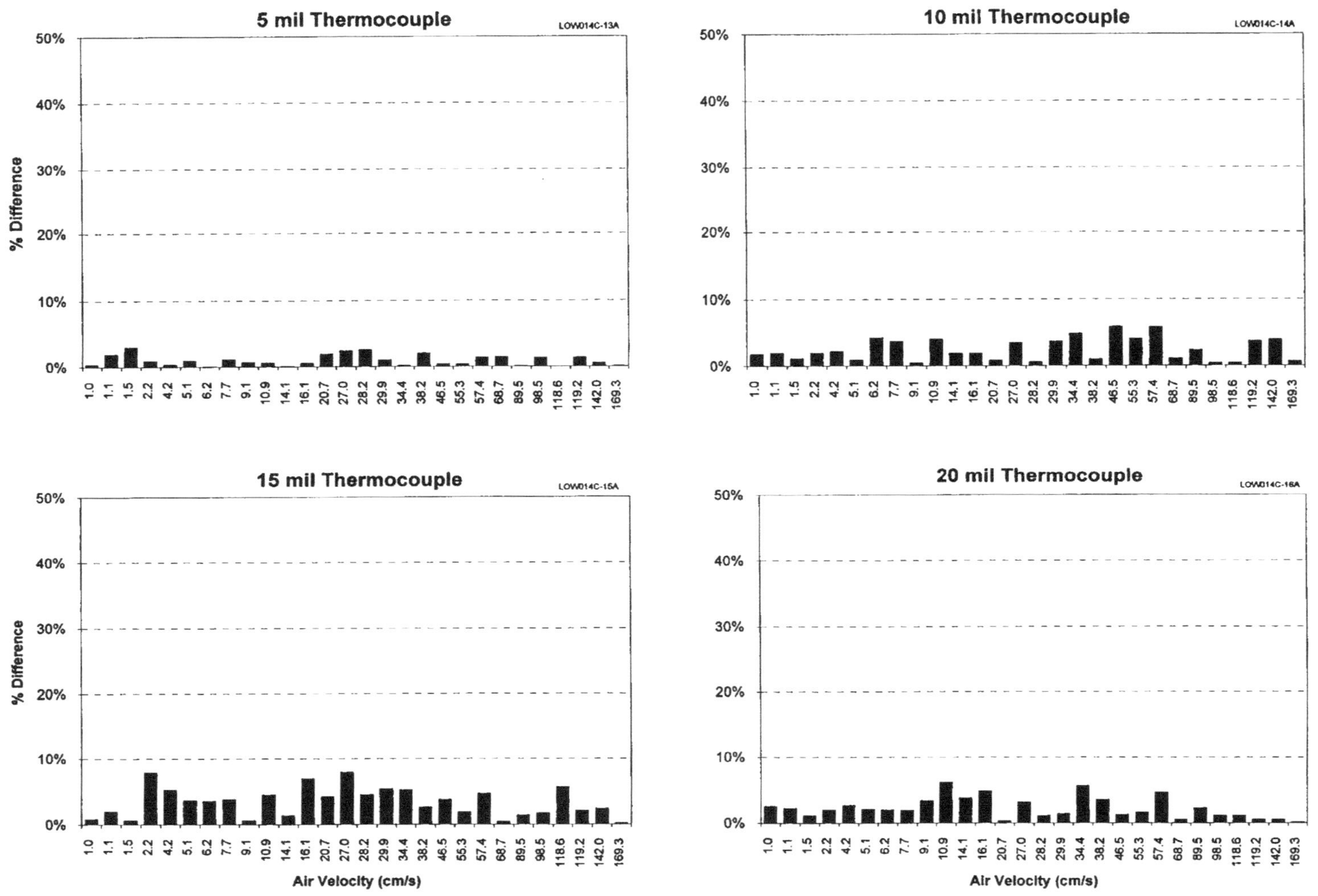

Figure 8.12 Errors Associated With Velocity Measurement Using the Series Equation

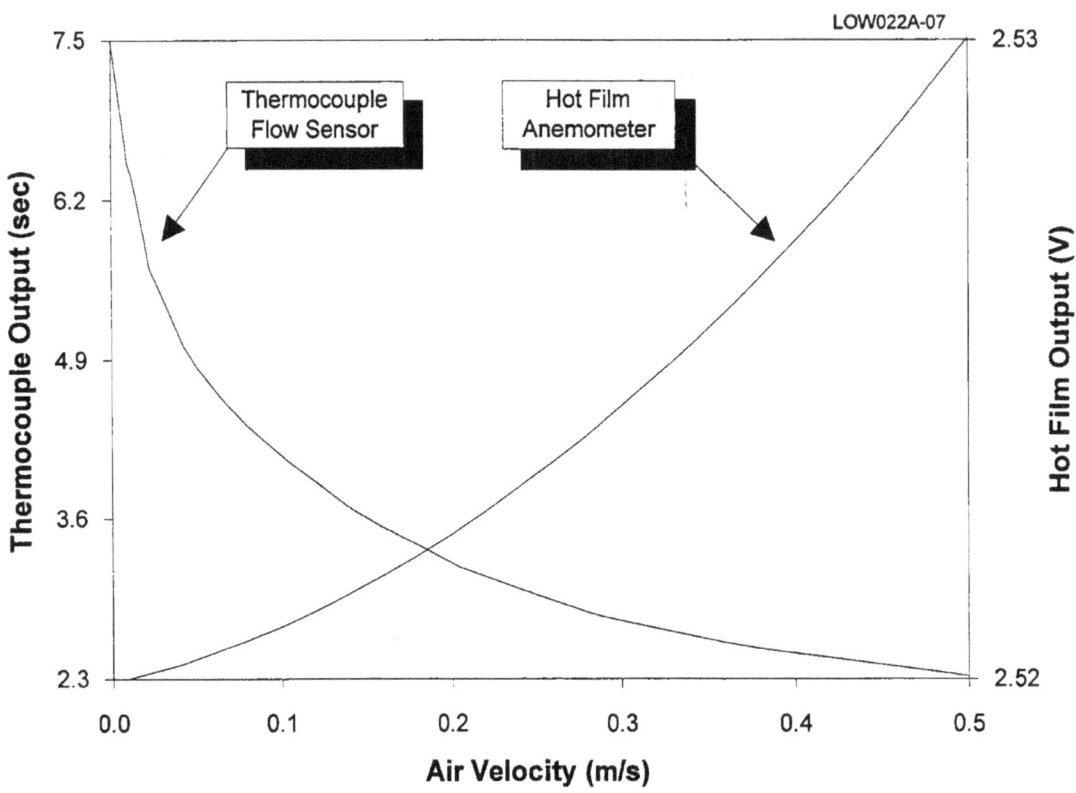

Figure 8.13 Comparison of a Hot Film Flow Sensor With
a 10 Mil Thermocouple Flow Sensor

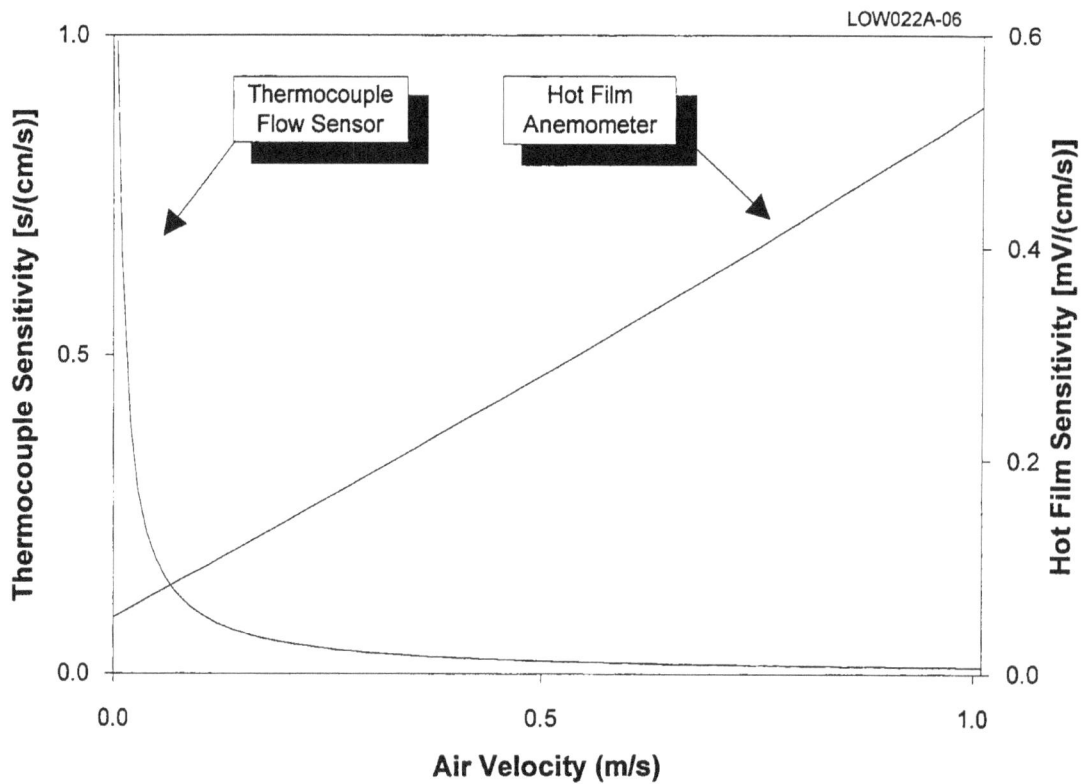

Figure 8.14 Sensitivity of Hot Film and 10 Mil Thermocouple
Sensors to Changes in Air Flow Rates

9.   ACCURACY AND SENSITIVITY OF THERMOCOUPLE FLOW SENSORS

The accuracy of thermocouple flow sensors depends on two factors:

1) Accuracy of the LCSR method for measurement of thermocouple response time, and

2) Accuracy of the response time-versus-flow correlation that is used to convert the response time results to flow velocity.

The average accuracy of the LCSR test is about 10 percent as indicated by the results given in section 8. The accuracy of response time-versus-flow correlations depends on the accuracy of the reference flow measurements which includes the accuracy and repeatability of the reference flow sensor, and the measurement precision. In this project, an LDV was used as the reference flow sensor. The LDV accuracy is 0.1 percent of measured velocity and its repeatability is excellent. Furthermore, very little precision error is involved in air flow measurement with an LDV. Therefore, the accuracy of the reference flow measurement is negligible compared to the LCSR accuracy of 10 percent. That is, the effective accuracy of thermocouple flow sensors is about 10 percent. The accuracy question will be addressed in more detail in Phase II.

Figure 9.1 shows sensitivity results for four thermocouple flow sensors of different sizes. These data are from response time-versus-flow rate tests performed using the LCSR method.

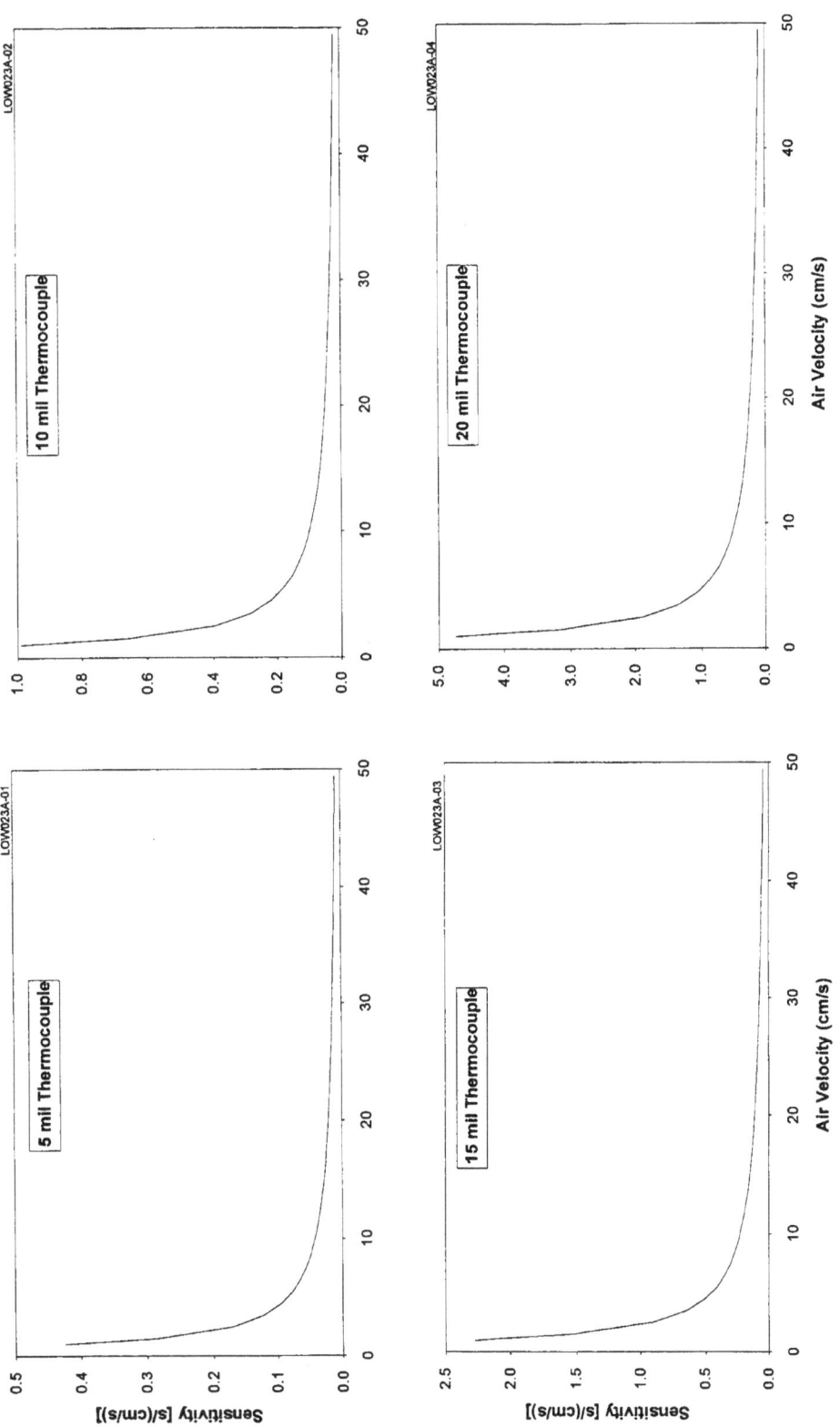

Figure 9.1 Sensitivity of Different Size Thermocouples

- 87 -

10. DETERMINING AIR FLOW PATTERNS

The Phase I effort was concerned only with demonstrating the feasibility of thermocouple flow sensors. However, some investigations were also conducted as to how these sensors may be used for determining flow direction in addition to flow velocity.

To measure flow direction, multiple thermocouple flow sensors may be used in the monitoring area and the results may be analyzed using existing empirical, physical, or mathematical models to estimate air flow patterns. A physical model developed based on Computational Fluid Dynamics (CFD) was identified during Phase I.[16] In Phase II, this model will be used with flow velocity and temperature data provided by thermocouple flow sensors to determine its potential for providing indoor air flow patterns. The

potential of neural networks will also be examined in Phase II. It is anticipated that neural network empirical models may be as effective as CFD and other physical models for determining air flow patterns. This point will be verified in Phase II.

Multiple thermocouples in a monitoring area could simply be multiplexed to a single LCSR test system programmed to give velocity and temperature data for use with theoretical models for estimating flow patterns. As such, only one LCSR data acquisition and data processing system would be sufficient for an almost unlimited number of flow sensors. Figure 10.1 illustrates the concept of using multiple thermocouples to define air flow velocity and direction.

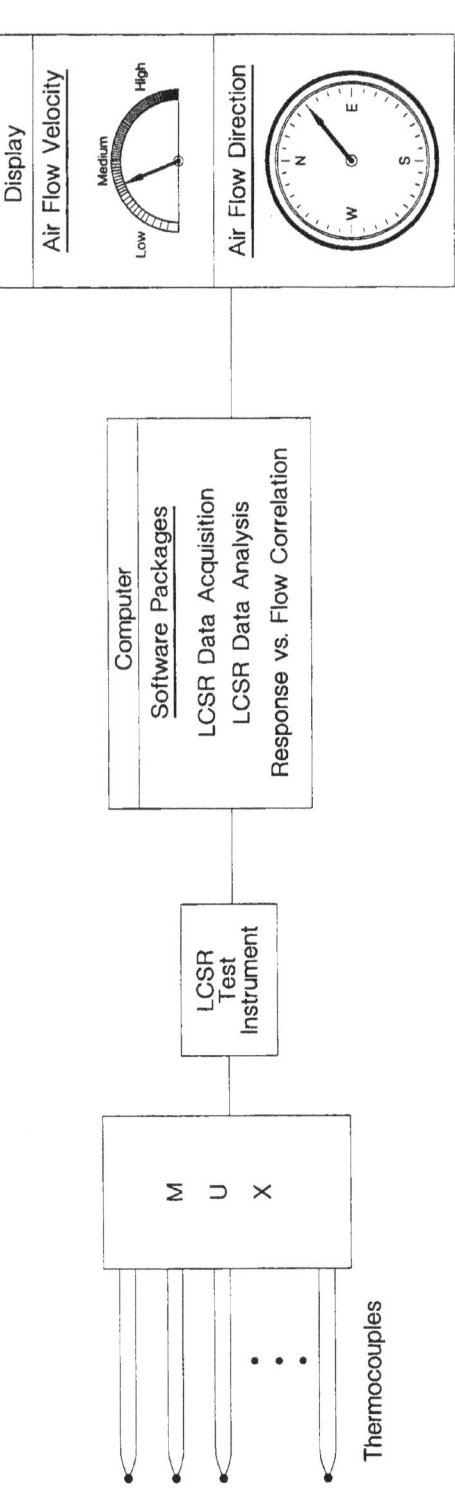

10.1 Multiple Thermocouple Flow Sensor for Measurement of Velocity and Direction

11. SURVEY OF POTENTIAL USERS

An informal survey was conducted to identify potential users of low flow sensors in nuclear and other industries and their specifications and requirements. The following facilities indicated an interest in the Phase I project and spent time discussing the issue with AMS, providing suggestions, or helping with the review of the project report.

- Four Nuclear Fuel Fabrication Facilities
- A Nuclear Waste Handling Facility
- An Environmental Consulting Firm
- An Engineering Consulting Firm
- Seven Nuclear Power Stations

Table 11.1 provides a listing of the key individuals contacted for the survey. This list is followed by Table 11.2 presenting the survey results.

11.1 Nuclear Fuel Fabrication Facilities

The Health Physics Department of the Fuel Fabrication Facilities at Westinghouse in Columbia, South Carolina, Babcock and Wilcox in Lynchburg, Virginia, ABB Combustion in Hematite, Missouri and Nuclear Fuel Services in Erwin, Tennessee were contacted and the following information was obtained:

Smoke candles and manual visualization techniques are currently used to perform air flow studies to identify the air flow patterns in general work areas where workers may be exposed to airborne radiation. This is done to establish a "breathing zone" as well as the

optimum locations to place air samplers. However, this technique is not desirable for an area where nuclear fuel products are developed because the residue left behind by the smoke necessitates protecting components. In addition, this study is normally performed when there is no activity in the area which may or may not represent the flow patterns present during actual working conditions.

The respondents to this survey favored the need to improve the existing methods or to develop a new air flow sensor which would provide not only information about the air flow patterns, but air velocity as well. The summary of their comments is given in Table 11.3.

11.2 Nuclear Waste Facility

The Scientific Ecology Group in Oak Ridge, Tennessee, which handles nuclear waste for the purpose of volume reduction, was visited and several observations were made about the air sampling and air balancing method they are currently using. Currently, air flow studies are performed to determine and verify that their facility has a negative air flow, relative to the outside environment. This is done by measuring air velocities at each opening in the building and by using smoke candles to identify the air flow patterns inside the facility. This process is labor intensive and requires many man-hours to perform. The proposed new sensor can reduce the amount of time required to perform the air flow studies and may also provide higher accuracies for the air flow measurements.

TABLE 11.1

Listing of Interested Parties Contacted in Phase I

Company	Contact
Nuclear Fuel Services Erwin, Tennessee	Mr. Andrew M. Maxin Vice President Mr. John Nagy Health Physics
Clayton Environmental Consulting Atlanta, Georgia	Ms. Alice Farrar Vice President Environmental Health Services
Scientific Ecology Group Oak Ridge, Tennessee	Mr. Michael Lauer Senior Radiological Engineer
Westinghouse Columbia, South Carolina	Mr. Jim Heath Health Physics
Advanced System Technology, Inc. Atlanta, Georgia	Mr. Wayne Knox President
Mechanical Engineering Department Texas A&M University College Station, Texas	Dr. A. R. McFarland Professor
Los Alamos National Laboratory Health Physics Measurement Group Los Alamos, New Mexico	Mr. Jeffrey J. Whicker Mr. John Rodgers Technical Staff

TABLE 11.2

Survey Summary

Type of Facility	Company Name	Tests Currently Used	Problems
Fuel Fabrication Facilities	ABB Combustion	Smoke tests periodically	Smoke residue
	Babcock & Wilcox	Smoke tests quarterly	Smoke residue and not enough smoke
	Nuclear Fuel Services	Smoke tests quarterly and comparison of fixed and lapel air samplers	None
	Westinghouse	Smoke tests quarterly	Smoke residue, not enough smoke, and unexplainable high particulate
Nuclear Waste Handling Facility	Scientific Ecology Group	Smoke tests, negative air balancing, velocity measurements at exits	Number of man-hours
Environmental Consulting Firm	Clayton Environmental Consulting	Indoor air quality control measurements	Devices not sensitive at low air flows
Nuclear Power Stations	Arkansas Nuclear One	Smoke candles and general sampling requirements	None
	D.C. Cook	Portable flowmeters and general sampling requirements	None
	Farley	Velocity and direction checked around fume hoods (1 in 3 yrs) and general sampling requirements	None
	Palisades	Lapel air samplers and general sampling requirements	None
	South Texas Project	Strings to visualize air flow direction and general sampling requirements	None
	Wolf Creek	Draeger smoke tubes and general sampling requirements	None
	Zion	General sampling requirements	None

TABLE 11.3

Fuel Fabrication Facilities Survey Results

Item	Comments
1	The fuel fabrication facilities surveyed used the smoke candle test method approximately 3-4 time per year.
2	The testing took 3-5 man-days to complete.
3	The air flow rates they are trying to determine range from 0.5 ft/sec (15 cm/s) to stagnant conditions.
4	All facilities expressed concerns about problems with the smoke tests. The major problem is that the smoke leaves a residue in "clean room" areas. It was also stated that not enough smoke is produced to accurately determine the air flow patterns.
5	Portable air samplers are used to sample for high concentrations of radioactive airborne particulate.
6	One problem related to this sampling, is that sometimes the particulate contamination would be high for no apparent reason. This would typically occur in the Recovery Operations Area.
7	All of these facilities said that they would be interested in purchasing the new flow sensor if the price was reasonable.

11.3 Environmental Consulting Firm

Clayton Environmental Consulting was contacted in Atlanta, Georgia, where they perform indoor air quality measurements for commercial buildings. The devices they are currently using are not sensitive at low air velocities. The proposed new sensor could be useful for the type of applications used in hospitals, particularly, in central supply facilities where ethylene oxide leaks are possible and difficult to track. In addition, the new sensor could aid in monitoring the indoor air quality in high-rise office buildings.

11.4 Engineering Consulting Firm

Advanced System Technology Inc. was contacted in Atlanta, Georgia, where they provide technical services in the area of health physics and environmental remediation to NRC, EPA, DOE, Department of Navy, and nuclear and non-nuclear facilities. They have performed extensive air flow characterization studies in a nuclear facility using smoke candles and visualization techniques. They provided some useful information about air flow pattern studies, and they strongly favor improvement in existing methods for air flow pattern studies in nuclear facilities. They have offered to provide suggestions and share their knowledge and expertise for development of an expert system for characterization of air flow pattern in nuclear facilities.

11.5 Nuclear Power Stations

Health physics departments of seven nuclear power stations were contacted to identify their air sampling practices and requirements. Most nuclear power stations do not use qualitative methods. They rely on engineering design features which include installation of air samplers around fume hoods, door ways, air supply and exhaust vents. Several plants require workers to wear lapel samplers in areas where airborne contamination may be present or there is a potential for radioactive release. Several personnel expressed that common sense and practical experience were used in the placement of air samplers. However, in some nuclear power stations qualitative studies are performed any time there is a change in the work environment which may have a potential of changing air flow patterns in the work area. This is done by using smoke devices for visualization of air flow patterns and assessment of air velocity in different locations within the work place. The plants that perform qualitative studies expressed strong support for the development of a new and easier method for air flow velocity and pattern identification. A summary of respondent comments is given in Table 11.4.

TABLE 11.4

Nuclear Power Stations Survey Results

Item	Comments
1	Nuclear power plants rarely use smoke candles to monitor air flows. They use design engineering specifications, practical experience, and common sense for placement of air samplers. Maintenance engineering will sometimes request the Health Physics Department to perform a smoke test.
2	All of the Health Physics engineers surveyed said that they only need a qualitative assessment of air flow direction to place their air samplers.
3	One nuclear power plant used a string method for determining air flow direction.
4	None of the nuclear power plants surveyed said that they had a need for a quantitative assessment of air velocity.
5	Most of the nuclear power plants surveyed have permanently installed air samplers around fume hoods, door ways, exhaust vents, and other work areas.
6	Most of the plants perform portable air sampling, where a rotameter pulls a specified volume of air through a filter collecting airborne particulates for later analysis to determine radioactive contamination and chemical content.
7	Some plants exclusively use lapel air samplers to determine personnel exposure to airborne contaminants.

12. CONCLUSIONS

The feasibility of a low flow sensor that uses a thermocouple as its sensing element was demonstrated in this project. The sensor can measure low flow rates in air and other gases with much better sensitivity than existing flow sensors. In addition, it has the potential to be used with theoretical models to yield information about air flow direction and help in determining indoor air flow patterns.

This project was initiated to develop a better means of establishing air flow patterns in radiation work areas. The information obtained using these improved methods will allow health physics personnel to more accurately determine the placement of air samplers in work areas for representative sampling of radioactive airborne contaminants.

REFERENCES

1. U.S. Nuclear Regulatory Commission, "Air Sampling In The Workplace," Regulatory Guide 8.25, Washington, D.C., 1992.

2. U.S. Nuclear Regulatory Commission, "Air Sampling In The Workplace," NUREG-1400, Washington, D.C., 1993.

3. Hashemian, H.M., "New Technology for Remote Testing of Response Time of Installed Thermocouples," Arnold Engineering Development Center AEDCTR-9-26, January 1992.

4. Hashemian, H.M., Petersen, K.M., "Loop Current Step Response Method for In-Place Measurement of Response Time of Installed RTDs and Thermocouples," Seventh International Symposium on Temperature, Its Measurement and Control in Science and Industry, American Institute of Physics, Vol. 6, Part 2, Toronto, Canada, April 1992.

5. Benedict, Robert P., Fundamentals of Temperature, Pressure And Flow Measurements 3rd Edition, Wiley, New York, New York, 1984.

6. Kurz, Jerry, "Characteristics And Applications Of Industrial Thermal Mass Flow Transmitters," 47th Annual Symposium on Instrumentation for Process Industries, ISA, 1992.

7. Wells, M.R., "3-D Laser Doppler Velocimeter Measurements Beneath A Drill Bit In A Cylindrical Test Cell," ISA Transactions, Vol. 28, pp. 509-528, 1989.

8. Miller, R.W., Flow Measurement Engineering Hand Book Second Edition, McGraw Hill, New York, New York, 1989.

9. Gougon-Durand, Sophie, "Vortex Anemometer For Measurement Of Both Speed And Direction Of The Wind," ISA Transactions, Vol. 32, pp. 393-396, 1993.

10. Hashemian, H.M., et al, "Time Response of Temperature Sensors," Paper C.I. 80-674, Instrument Society of America, International Conference and Exhibit, Houston, Texas, October 1980.

11. Rohsenow, W.M., Choi, H.Y., Heat Mass And Momentum Transfer, Prentice-Hall, Englewood Cliffs, New Jersey, 1961.

12. Holman, J.P., Heat Transfer 4th Edition, McGraw Hill, New York, New York 1976.

13. Carroll, R.M., Shepard, R.L., "Measurement of Transient Response of Thermocouples and Resistance Thermometers Using an In-Situ Method," Oak Ridge National Laboratory, Report Number ORNL/TM-4573, Oak Ridge, Tennessee, June 1977.

REFERENCES
(Continued)

14.   Shepard, R.L., Carroll, R.M., "Thermocouple Response Time Measurement and Improvement." Notes on short course entitled "Sensor Response Time Testing in Nuclear Power Plants," College of Engineering, The University of Tennessee, Knoxville, Tennessee, June 1977.

15.   Kerlin, T.W., "Analytical Methods for Interpreting In-Situ Measurements of Response Times in Thermocouples and Resistance Thermometers," Oak Ridge National Laboratory, Report Number ORNL/TM-4912, Oak Ridge, Tennessee, March 1976.

16.   Baker, A.J. Kelso, R.M., Roy, S., "CFD Experiment Characterization of Airborne Contaminant Transport for Two Practical 3D Room Air Flow Fields," Building and Environment, Vol. 29, No. 3, 1994.

APPENDIX A

GENERAL PROCEDURE FOR LCSR
TESTING OF THERMOCOUPLES

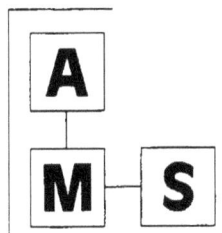

ANALYSIS AND
MEASUREMENT SERVICES
CORPORATION

AMS 9111 CROSS PARK DRIVE / KNOXVILLE, TN 37923 USA (615) 691-1756

Procedure # TCT8801R0

THERMOCOUPLE LOOP CURRENT STEP
RESPONSE TEST PROCEDURE

June 1988

Procedure #: <u>TCT8801R0</u>

Submitted By: _____ Date: _____6/21/88_____

Reviewed By: _____Dan D. Bewly_____ Date: _____6/22/88_____

Approved By: _____H. M. Aashemian_____ Date: _____6/22/88_____

1. PURPOSE

The purpose of this procedure is to give detailed instructions for in-situ response time testing of Thermocouples (TCs) using the Loop Current Step Response (LCSR) method. This procedure is based on using the AMS Model ETC-1 Instrument along with general purpose amplifiers, filters and data acquisition equipment.

The user of this procedure is assumed to be familiar with the LCSR methods.

2. PRINCIPLE OF THE TEST

The LCSR test may be performed by heating the TC with an applied AC or DC current for a given period of time, removing the current, and observing the TC output voltage as it cools back down to the process temperature. The resulting output is a transient that can be analyzed to give the response time of the TC as it is installed. This procedure does not address the analysis of the LCSR data. This procedure covers the performance of the LCSR test and the acquisition of the data.

3. TEST EQUIPMENT

The following equipment is required for performing the LCSR test:

1. AMS Model ETC-1 Thermocouple LCSR Test Instrument.

2. Strip chart recorder (dual channel).

3. Digital multimeter (DMM).

4. Amplifier(s) (Optional).

5. Filter (Optional).

6. Spare TC for equipment check-out.

7. Data acquisition equipment.

8. Log book with data sheets and graph paper.

Items 3, 4, 5 and 7 shall have valid calibrations traceable to NBS

TCT8801R0

4. PRECAUTIONS

The ETC-1 is capable of applying hazardous voltages to the TC being tested. These voltages are also present on the front panel of the instrument. The user should exercise care in connecting the TC to the ETC-1. The user should also understand that misuse of the equipment could potentially damage TCs being tested.

5. PROCESS CONDITIONS FOR THE TEST

The LCSR test must be performed at steady state conditions. That is, the process temperature, pressure and flow must be as constant as possible during a LCSR test. The inherent random fluctuations of these parameters may be tolerated but should be minimized by:

1. Using sufficient heating current in the LCSR test to improve the signal to noise ratio.

2. Using sufficient heating time to obtain proper signal to noise ratio.

3. Taking more than one data set on each sensor, then using an averaging technique to smooth the data. Depending on the amount and magnitude of the fluctuations, from 5 to 20 data sets are usually sufficient for this averaging.

The time constant of a TC is a function of fluid temperature and flow rate of the fluid around the sensor. Therefore, a plant TC must be tested at or around the normal operating condition to yield the time constant of the sensor in-service.

6.0 TEST PROCEDURE

This section gives the step-by-step instructions for performing a LCSR test on an installed TC. The TC must be disconnected from the plant signal conditioning equipment and the TC extension wires themselves will be connected to the ETC-1.

6.1 Equipment Checkout

6.1.1   Set up the equipment as near as possible to the cabinet where the TC wires are connected to the plant signal conditioning equipment.

6.1.2   CAUTION: Prior to connecting power to the equipment, make sure that all power switches are OFF, and that the ETC-1 CONTROL switch is in the RESET position.

6.1.3   Connect the equipment to 115 VAC power source, turn power switches ON and allow 15 minutes warm-up time.

6.1.4   Connect the output of the ETC-1, through the appropriate signal conditioning equipment (if any), to one channel of the strip chart recorder.

6.1.5   Connect the DMM to the same signal as the strip chart recorder (signal conditioning equipment output or ETC-1 output as appropriate).

6.1.6   Ensure that the ETC-1 VOLTAGE ADJUSTMENT is set to 0% (full counterclockwise) and that the CONTROL switch is set to RESET.

6.1.7   Connect the ETC-1 CONTROL OUT BNC connector to the second channel of the strip chart recorder.

6.1.8   Adjust the strip chart recorder channel connected to signal output to a full scale voltage that matches that of the data acquisition equipment.

6.1.9   Connect the spare TC to the reference junction on the front panel of the ETC-1.

6.1.10 Set the CURRENT CONTROL SWITCH on the ETC-1 to the ENABLE position and increase the VOLTAGE ADJUSTMENT to achieve the desired heating current while observing the heating current AMMETER on the ETC-1.

6.1.11 Set the CURRENT CONTROL SWITCH to the RESET position. Wait for the TC to return to ambient temperature.

6.1.12 Start the STRIP chart recorder at an appropriate speed.

6.1.13 Set the CURRENT CONTROL SWITCH to the ENABLE position. Wait for the desired heating time.

6.1.14 Set the CURRENT CONTROL SWITCH to the RESET position. Observe the cooling transient on the strip chart recorder.

6.1.15 Adjust the amplifier gain and offset and the filter as necessary to obtain a transient scaled correctly for the data acquisition equipment.

6.1.16 Repeat 6.1.13 to 6.1.15 as necessary to obtain a properly scaled transient.

6.1.17 Stop the strip chart recorder.

6.1.18 Ensure that the CURRENT CONTROL SWITCH is in the RESET position.

6.1.19 Turn the VOLTAGE ADJUSTMENT to 0% (full counterclockwise).

6.1.20 Disconnect the spare TC from the ETC-1.

6.1.21 Preserve the strip chart traces as proof of the equipment checkout.

<u>Acceptance Criteria</u>

The output obtained at step 6.1.16 must be reasonably clean (free of excessive high and low frequency noise) exponential transient.

<u>6.2 LCSR Test Procedure</u>

6.2.1 Perform the pre-testing procedure outlined in the above section. Do not connect a plant TC to the ETC-1 unless the procedure in Section 6.1 is satisfactorily completed and the acceptance criteria are met. Proper operation of the ETC-1 must be assured before any connection to any plant TC is made.

6.2.2 Ensure that the ETC-1 VOLTAGE ADJUSTMENT is set to 0% (Full counterclockwise) and that the CURRENT CONTROL SWITCH is set to RESET.

6.2.3 Measure the thermocouple loop resistance and record the value on the attached data sheet along with the other information requested in data sheet.

6.2.4 Connect the TC to be tested to the reference junction on the front panel of the ETC-1.

6.2.5 Record appropriate data on the attached data sheet.

TCT8801R0

6.2.6   Set the CURRENT CONTROL SWITCH on the ETC-1 to the ENABLE position and increase the VOLTAGE ADJUSTMENT to achieve the desired heating current while observing the heating current AMMETER on the ETC-1.

6.2.7   Set the CURRENT CONTROL SWITCH to the RESET position. Wait for the TC to return to ambient temperature.

6.2.8   Start the STRIP chart recorder at an appropriate speed.

6.2.9   Set the CURRENT CONTROL SWITCH to the ENABLE position. Wait for the desired heating time.

6.2.10   Set the CURRENT CONTROL SWITCH to the RESET position. Observe the cooling transient on the strip chart recorder.

6.2.11   Adjust the amplifier gain and offset and the filter as necessary to obtain a transient scaled correctly for the data acquisition equipment.

6.2.12   Repeat 6.2.9 to 6.2.11 as necessary to obtain a properly scaled transient.

6.2.13   Record appropriate data on the attached data sheet.

6.2.14   Connect the output of the signal conditioning equipment to the analog input of the data acquisition equipment.

6.2.15   Connect the CONTROL OUT signal from the ETC-1 to the trigger input of the data acquisition equipment.

6.2.16   Start the data acquisition equipment.

6.2.17   Set the CURRENT CONTROL SWITCH to the ENABLE position. Wait for the desired heating time.

6.2.18   Set the CURRENT CONTROL SWITCH to the RESET position. Ensure that the data acquisition equipment is recording the TC cooling transient.

6.2.19   Repeat 6.2.17 and 6.2.18 as necessary to obtain the desired number of LCSR transients.

6.2.20   Stop the data acquisition equipment.

6.2.21   Record appropriate data on the attached data sheet and sign the data sheet.

6.2.22   Turn the VOLTAGE ADJUSTMENT to 0% (full counterclockwise).

6.2.23   Ensure that the CURRENT CONTROL SWITCH is in the RESET position.

6.2.24   Disconnect the TC from the ETC-1.

Acceptance Criteria

The output obtained at step 6.2.18 must be reasonably clean (free of excessive high and low frequency noise) exponential transient.

6.2.25   Repeat 6.2.3 to 6.2.24 for each TC to be tested.

THERMOCOUPLE LCSR DATA SHEET

Date _____   Time _____

Plant _____

Sensor ID _____ Manufacturer_____

Model # _____

Type: _____ E _____ J _____ K _____ T _____ Other

Wire Length _____   Wire Diameter (Gage) _____

Junction Type: _____ Exposed   _____ Sheathed   _____ Sheathed
                         Junction   (Insulated   (Grounded
                                                 Junction)   Junction)

Thermocouple Loop Resistance _____ (Ohms)

Installation Remarks: _____

_____

_____

Process   Flow _____

Conditions: Temperature _____

             Service _____

             Remarks _____

             _____

Test Conditions:

    Heating Current _____ (Amps)   Heating Time _____ (Sec)

    Output Voltage _____ (Volts)

    Filter Setting(s) _____

    Amplifier Gain(s) _____

Data Recording:

    Disk ID _____   File Names: _____ To _____

    Delta T _____   Number of Samples _____

Remarks: _____

_____

_____

_____

Signature

TCT8801R0   Page 7 of 7

- A-9 -

APPENDIX B

RAW LCSR AND PLUNGE
TEST TRANSIENTS

APPENDIX B

Raw LCSR and Plunge
Test Transients

This appendix contains samples of the LCSR and plunge test transients generated in the course of the Phase I project. Table B1 gives the tag number and description of the thermocouples for which raw data is provided in this Appendix. Raw data is shown based on thermocouple sizes. For each size, four LCSR transients are shown. The LCSR transients are followed by plunge tests transients. For each size, one typical plunge test transient is included at the end of this Appendix.

TABLE B.1

Listing of Thermocouples Used in This Project

Item	Wire Diameter	Tag No.	Loop Resistance (ohms)
1		N05A	24.6
2	5 mil	N05B	23.1
3		N05C	23.7
4		N05D	22.2
5		N10A	8.7
6	10 mil	N10B	7.7
7		N10C	8.6
8		N10D	9.5
9		N15A	4.9
10	15 mil	N15B	4.9
11		N15C	5.0
12		N15D	4.7
13		N20A	1.5
14	20 mil	N20B	1.3
15		N20C	1.3
16		N20D	1.6

Note: mil = 1/1000 inch
         All thermocouples listed above are Type K exposed junction

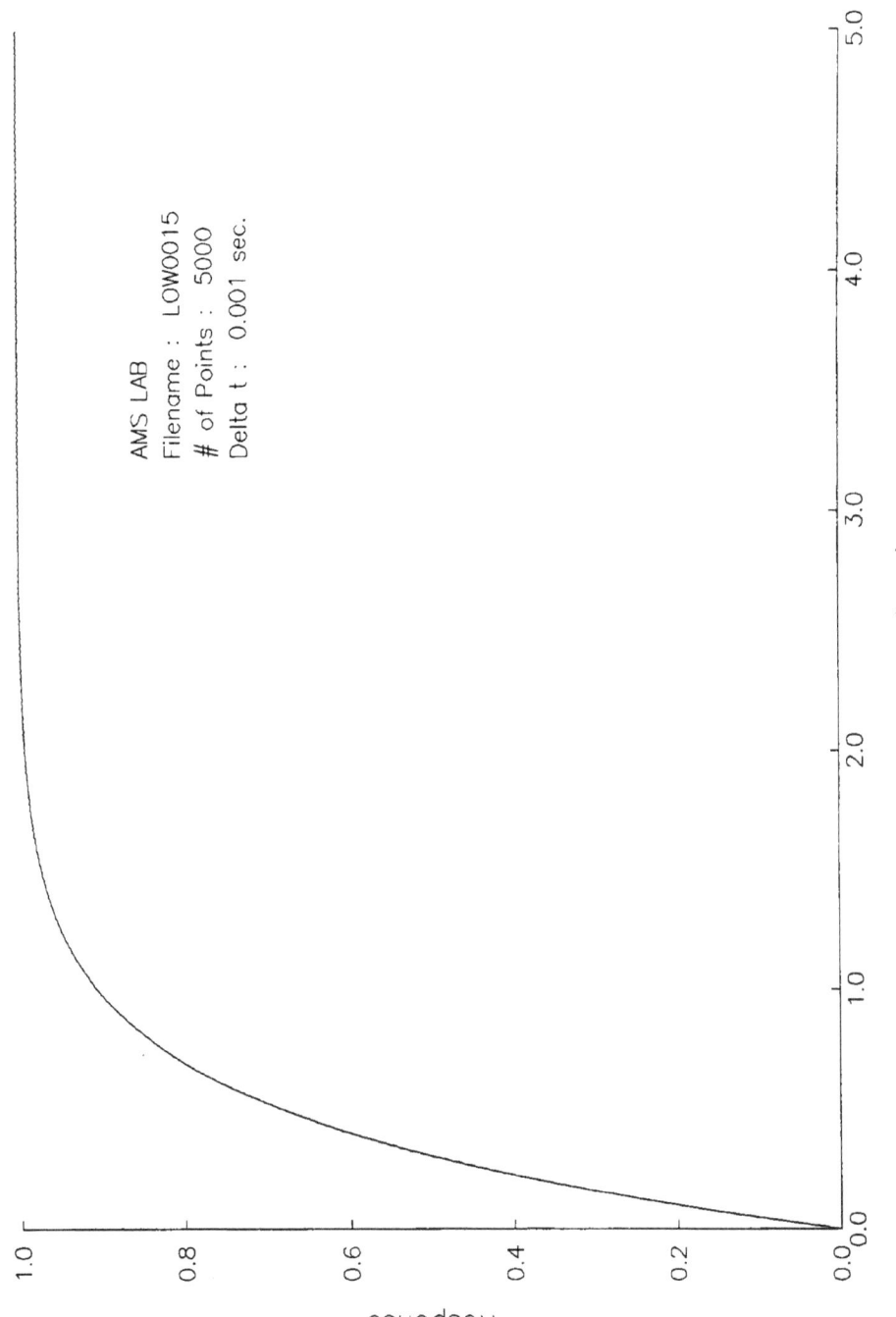

AMS LAB
Filename : LOW0015
# of Points : 5000
Delta t : 0.001 sec.

Figure B1. Raw LCSR Transient for Sensor Tag No. N05A .

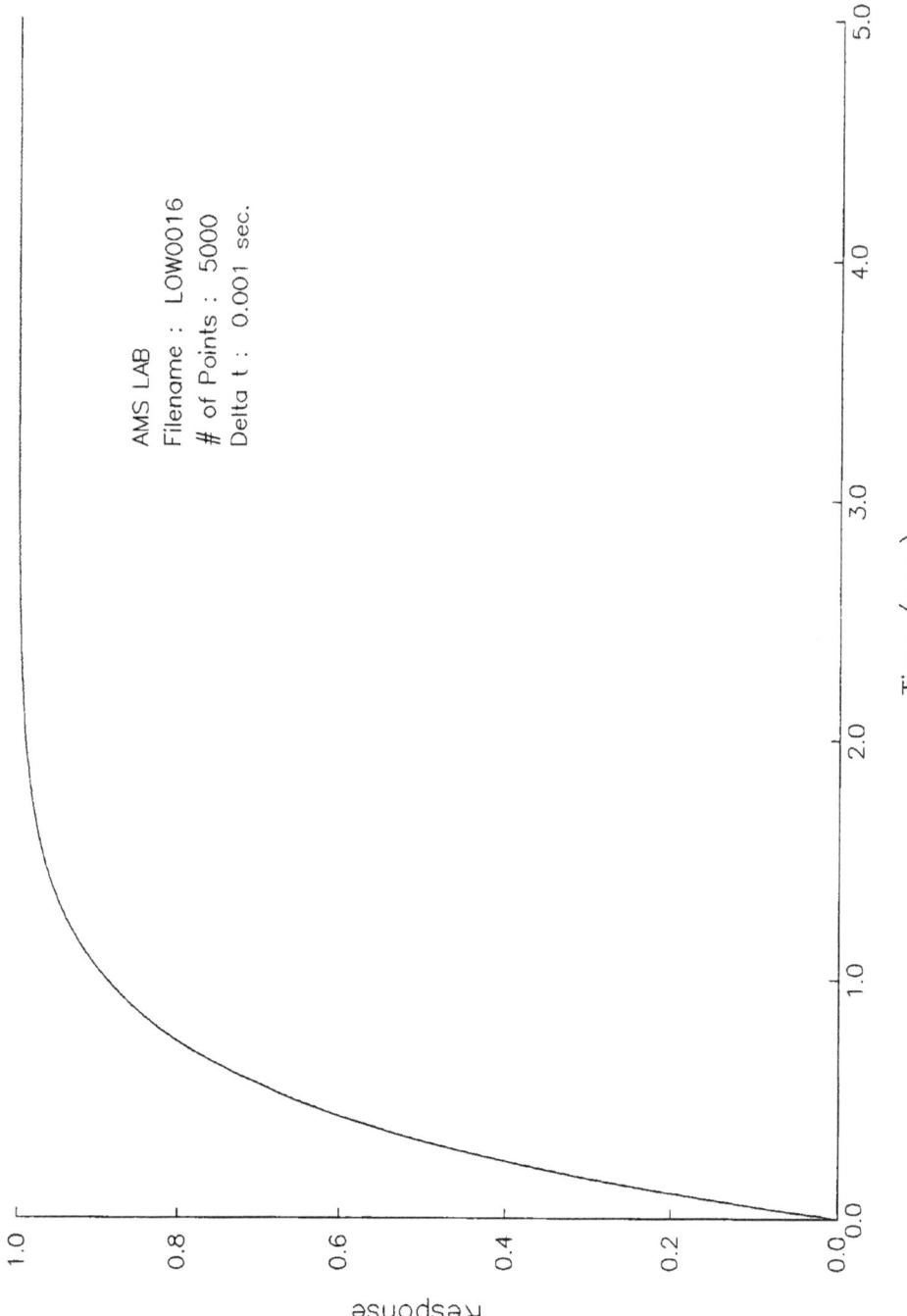

AMS LAB
Filename : LOW0016
# of Points : 5000
Delta t : 0.001 sec.

Time (sec.)

Response

Figure B2.   Raw LCSR Transient for Sensor Tag No. N05B .

Page B2

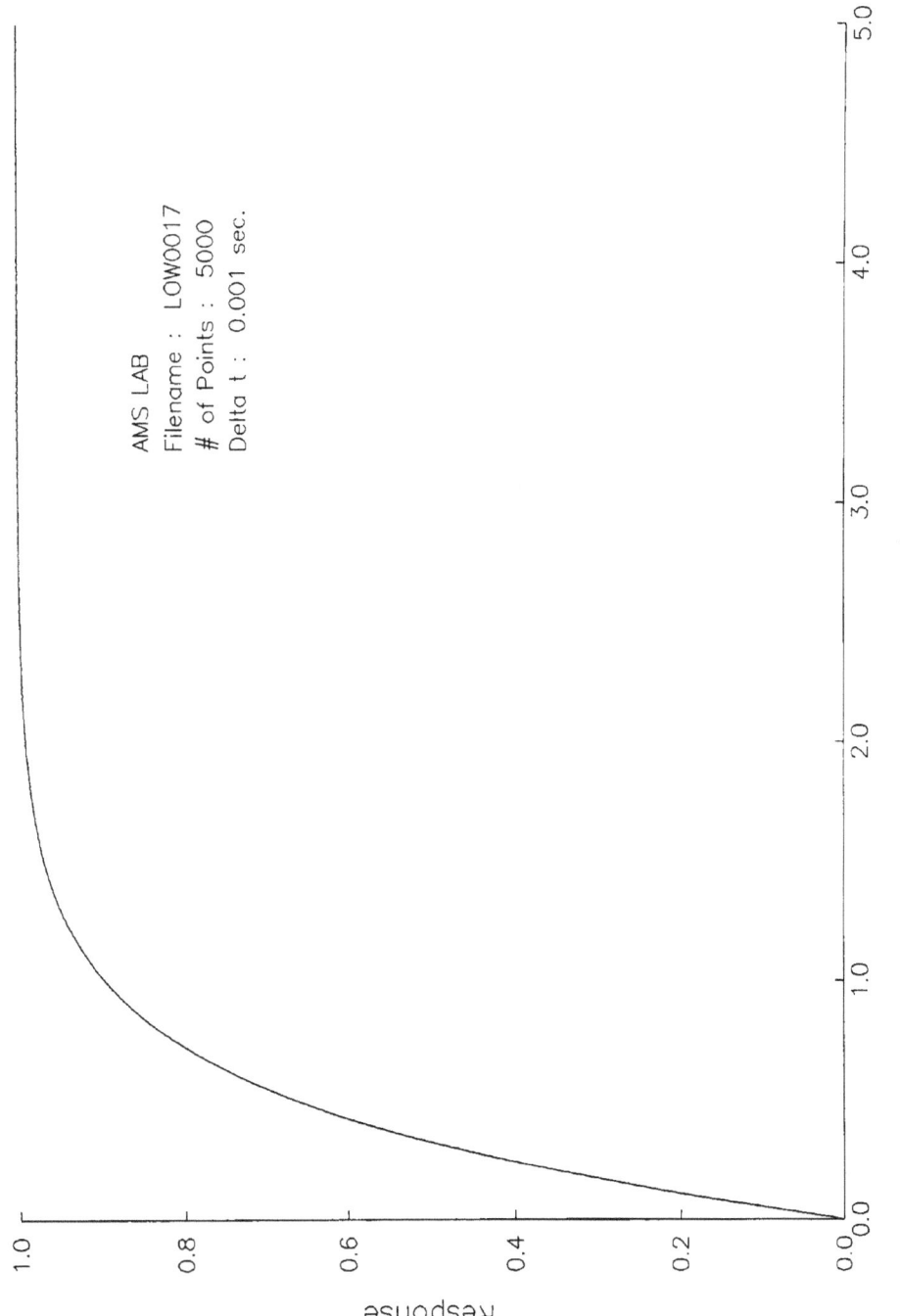

AMS LAB
Filename : LOW0017
# of Points : 5000
Delta t : 0.001 sec.

Time (sec.)

Response

Figure B3.   Raw LCSR Transient for Sensor Tag No. N05C .

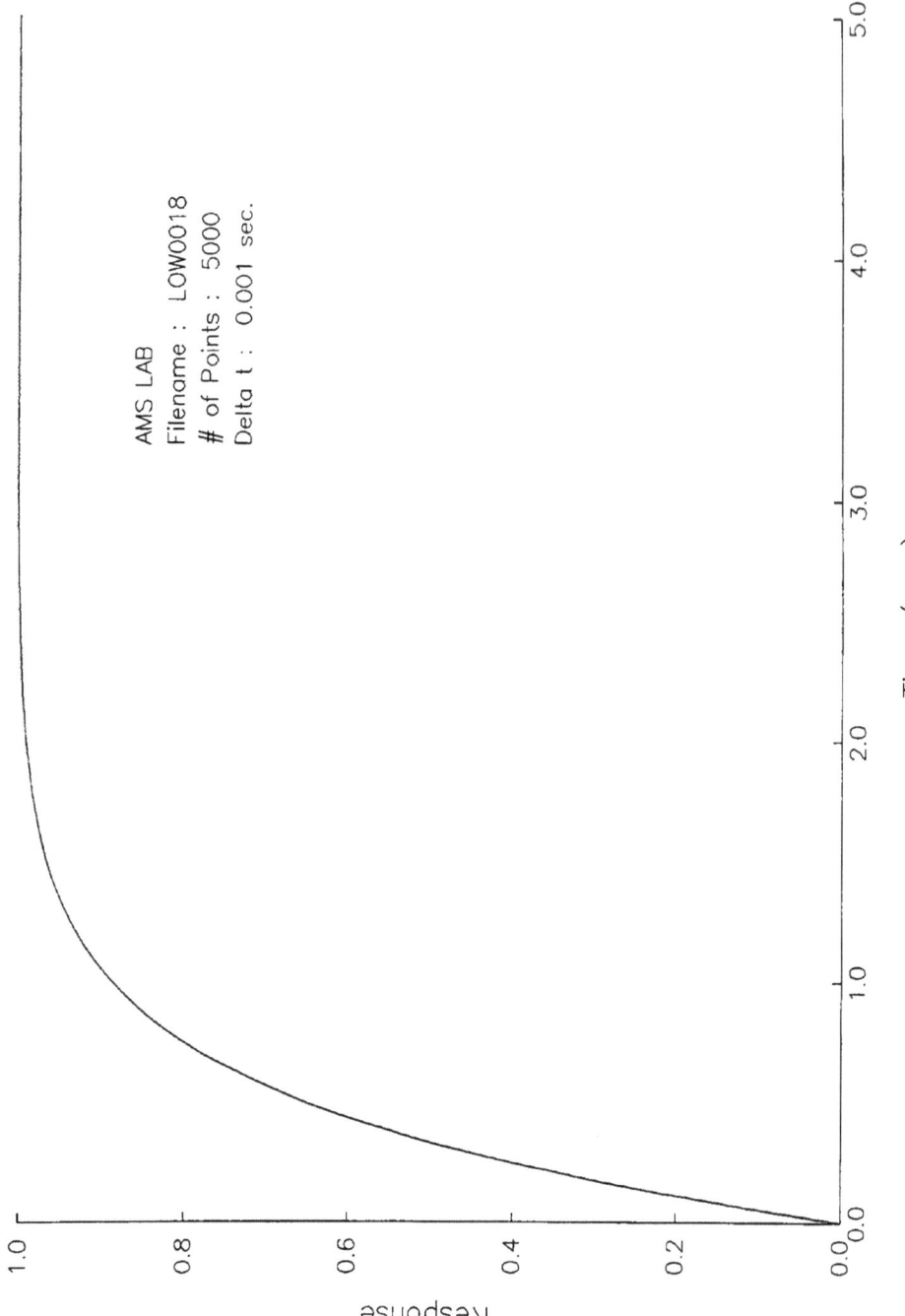

AMS LAB
Filename : LOW0018
# of Points : 5000
Delta t : 0.001 sec.

Response

Time (sec.)

Figure B4.   Raw LCSR Transient for Sensor Tag No. N05D .

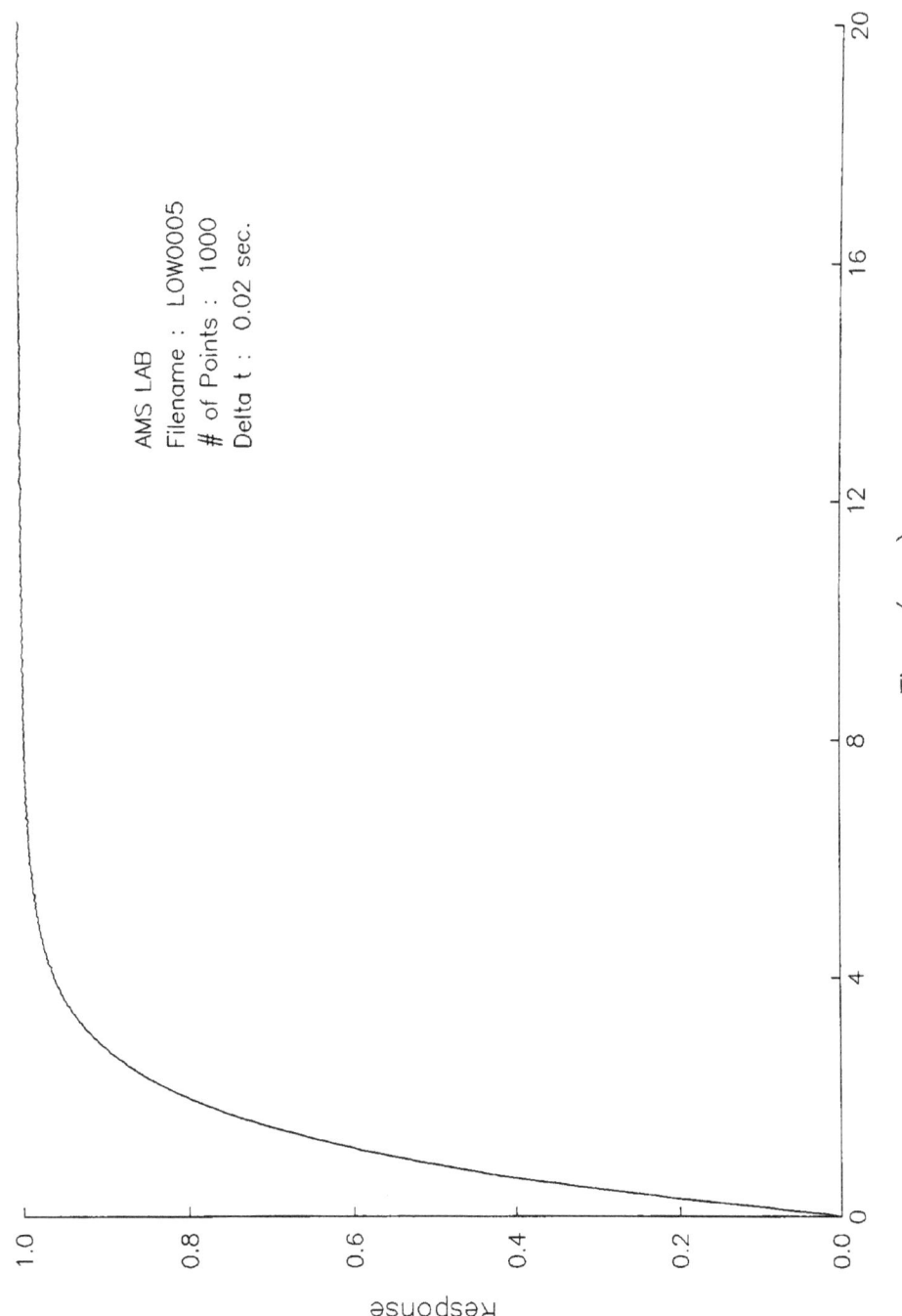

AMS LAB
Filename : LOW0005
# of Points : 1000
Delta t : 0.02 sec.

Time (sec.)

Response

Figure B5.   Raw LCSR Transient for Sensor Tag No. N10A .

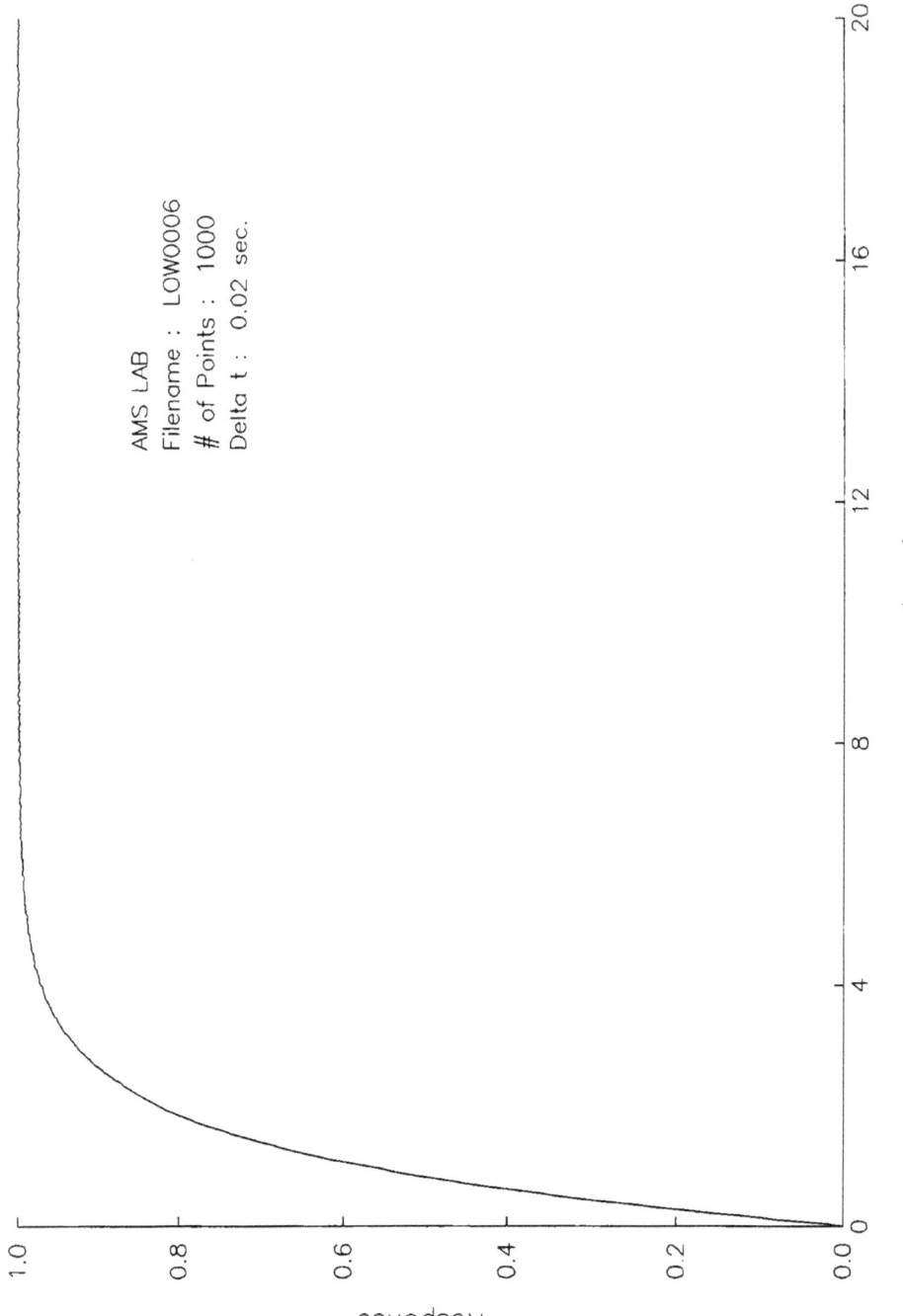

AMS LAB
Filename : LOW0006
# of Points : 1000
Delta t : 0.02 sec.

Figure B6.   Raw LCSR Transient for Sensor Tag No. N10B .

Page B6

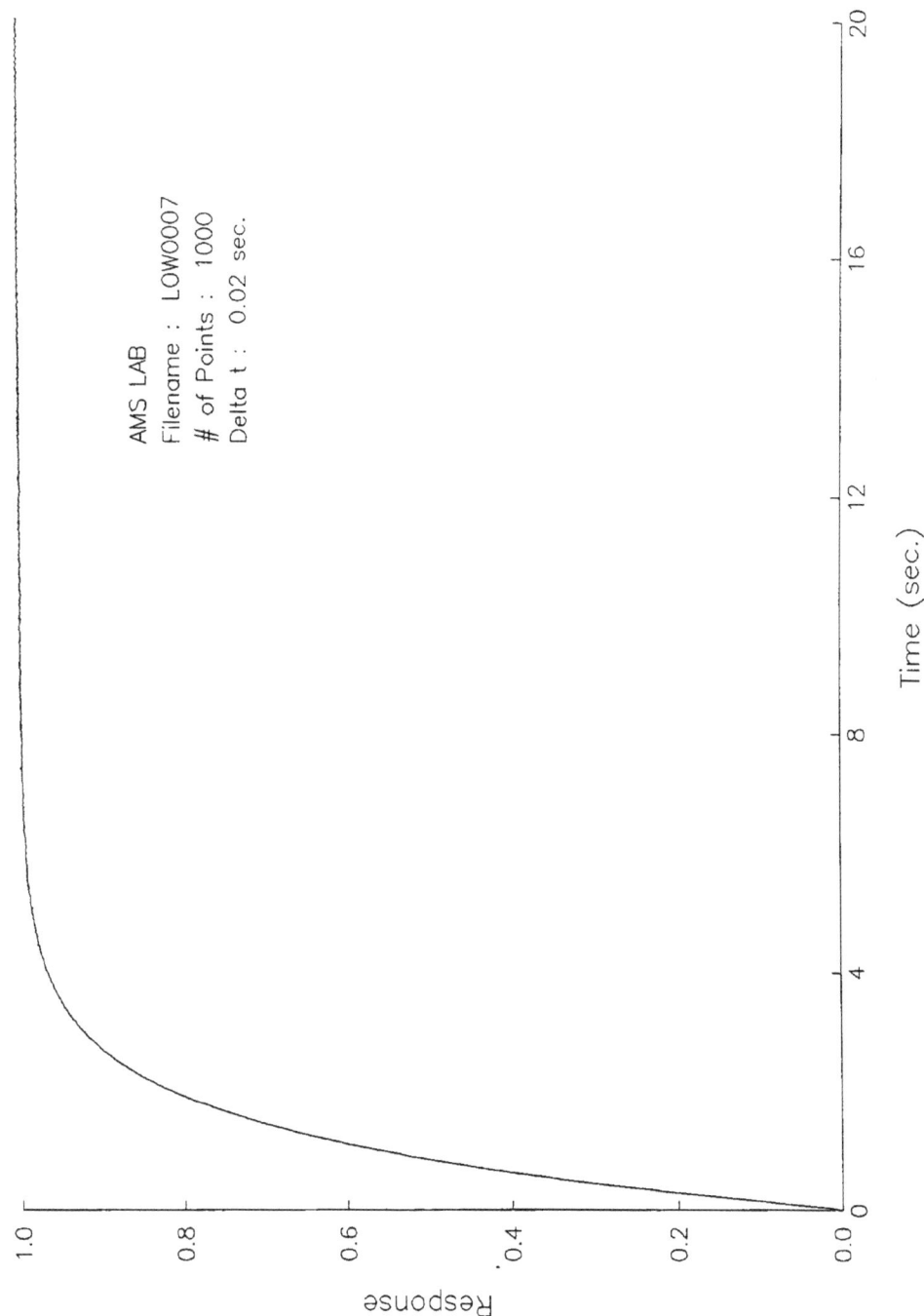

AMS LAB
Filename : LOW0007
# of Points : 1000
Delta t : 0.02 sec.

Time (sec.)

Response

Figure B7.   Raw LCSR Transient for Sensor Tag No. N10C .

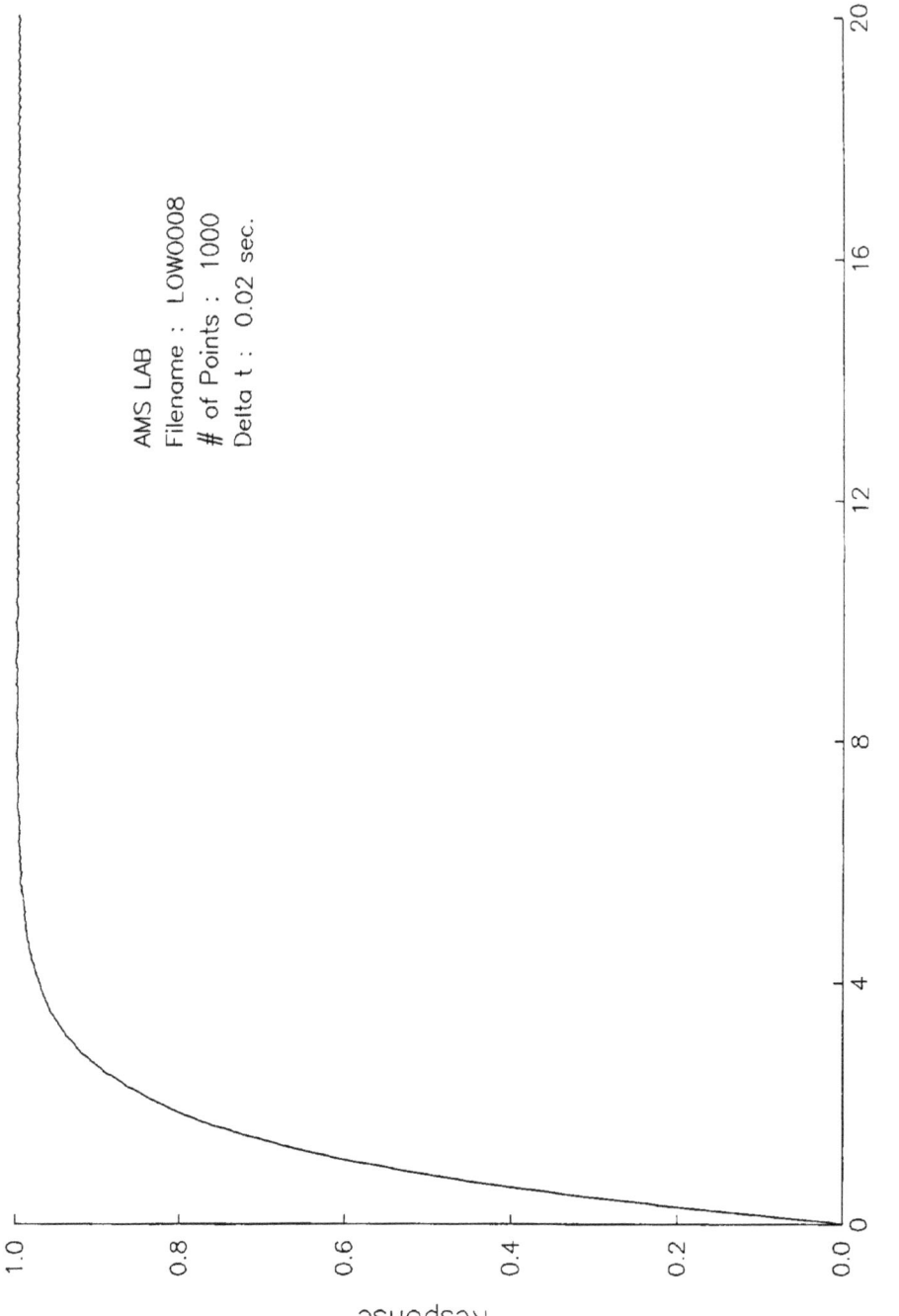

AMS LAB
Filename : LOW0008
# of Points : 1000
Delta t : 0.02 sec.

Response

Time (sec.)

Figure B8.   Raw LCSR Transient for Sensor Tag No. N10D .

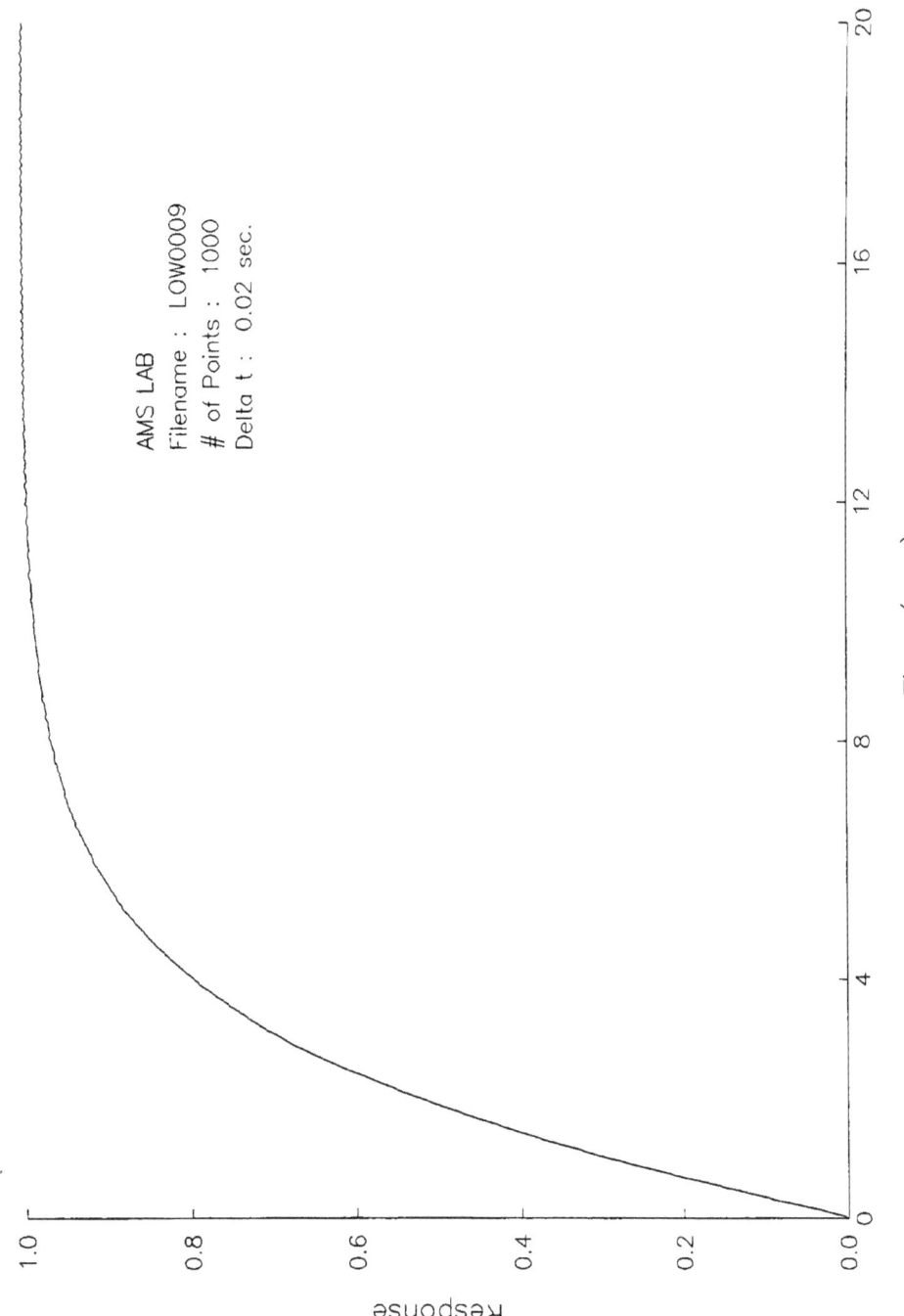

AMS LAB
Filename : LOW0009
# of Points : 1000
Delta t : 0.02 sec.

Figure B9.   Raw LCSR Transient for Sensor Tag No. N15A .

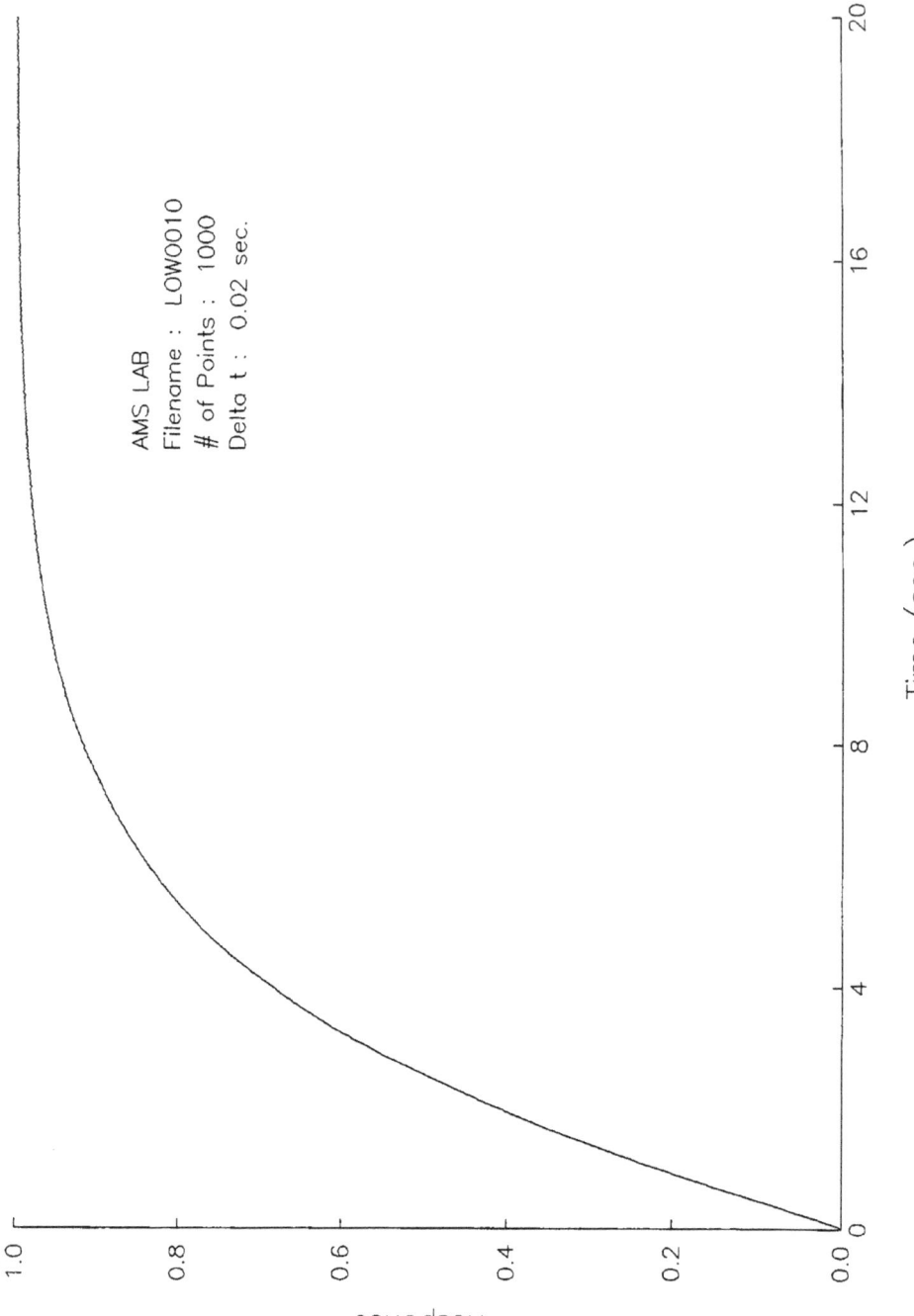

AMS LAB
Filename : LOW0010
# of Points : 1000
Delta t : 0.02 sec.

Figure B10.   Raw LCSR Transient for Sensor Tag No. N15B .

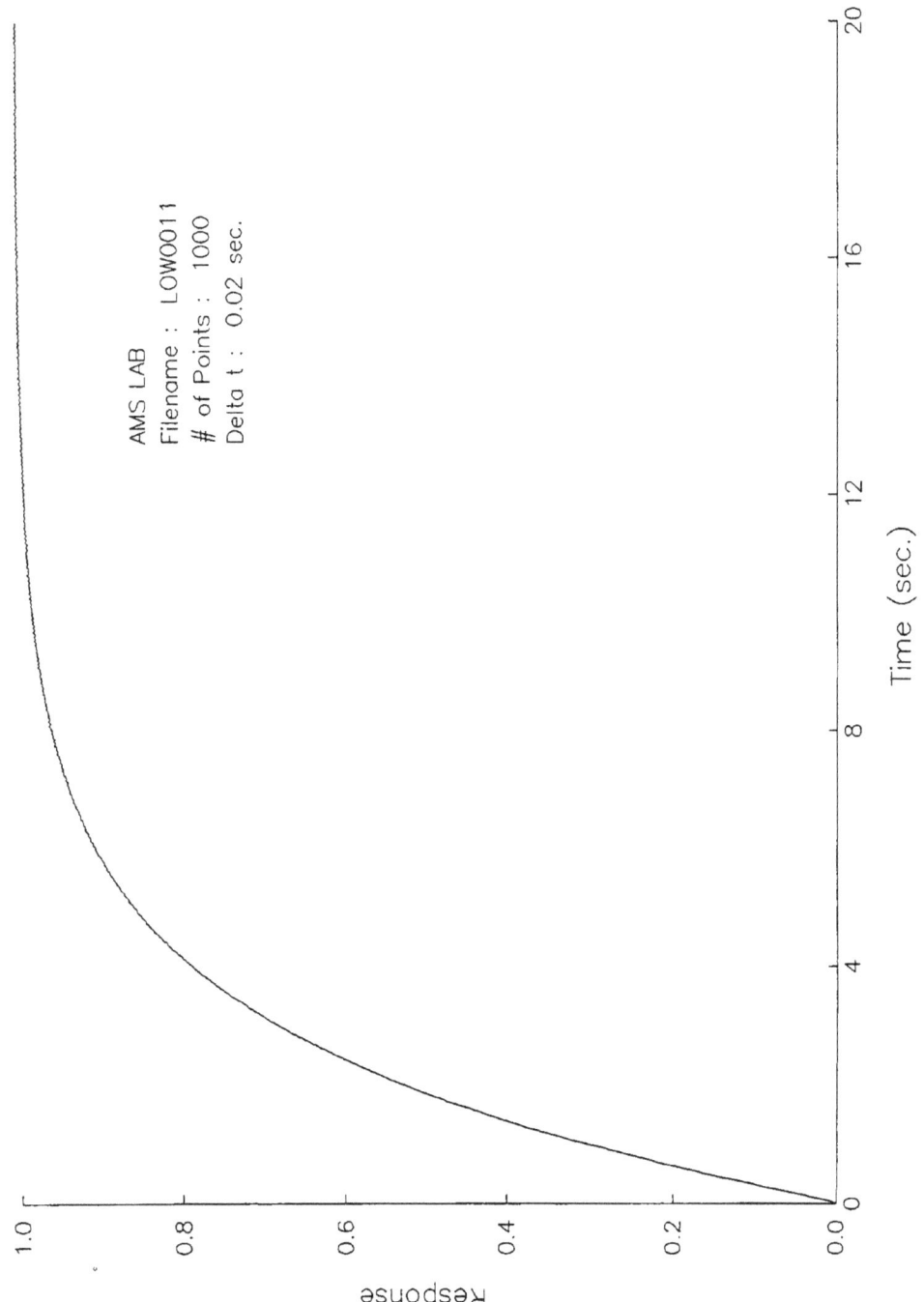

AMS LAB
Filename : LOW0011
# of Points : 1000
Delta t : 0.02 sec.

Time (sec.)

Response

Figure B11.   Raw LCSR Transient for Sensor Tag No. N15C .

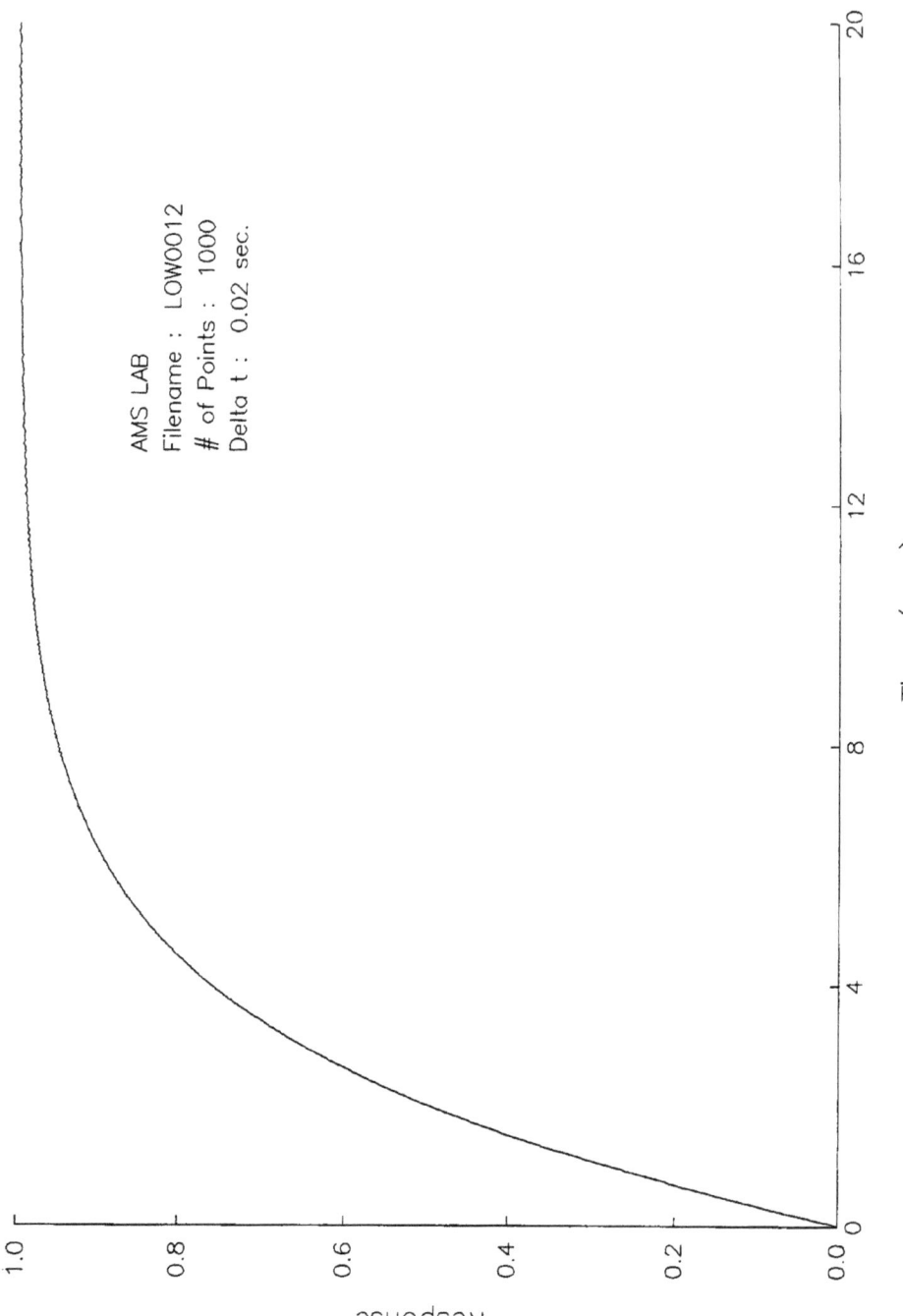

Figure B12.   Raw LCSR Transient for Sensor Tag No. N15D .

Page B12

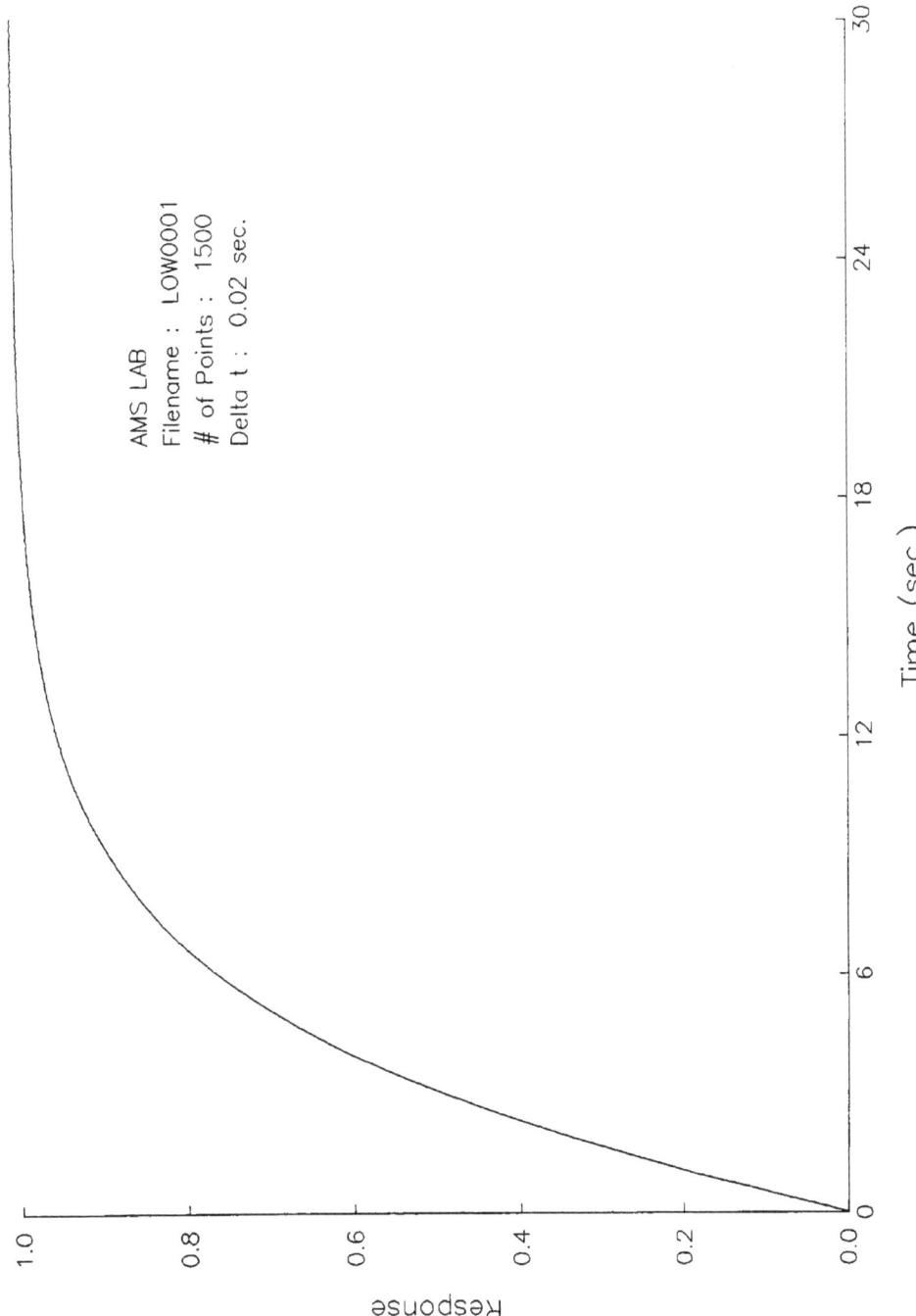

AMS LAB
Filename : LOW0001
# of Points : 1500
Delta t : 0.02 sec.

Figure B13. Raw LCSR Transient for Sensor Tag No. N20A .

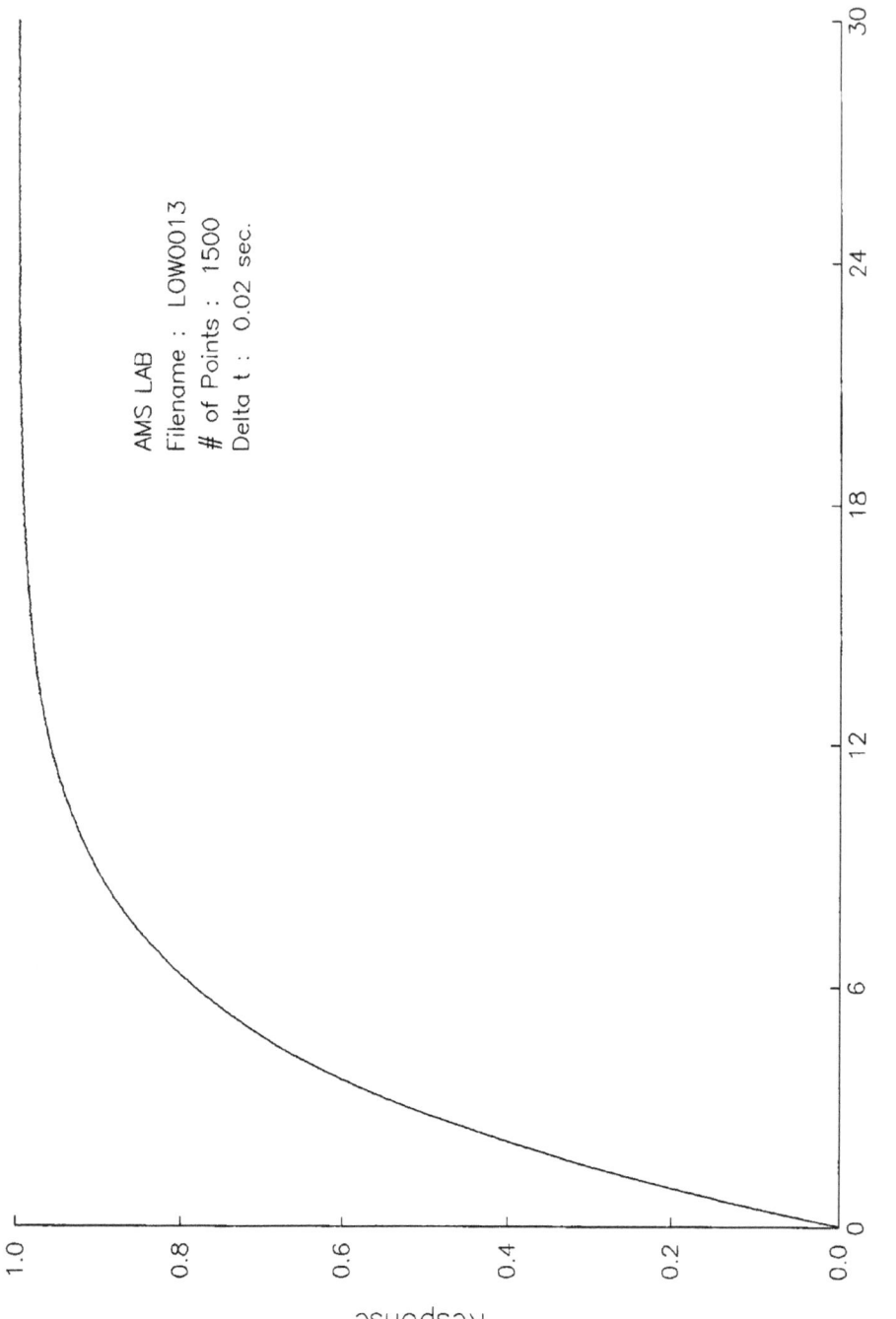

AMS LAB
Filename : LOW0013
# of Points : 1500
Delta t : 0.02 sec.

Response

Time (sec.)

Figure B14.   Raw LCSR Transient for Sensor Tag No. N20B .

Page B14

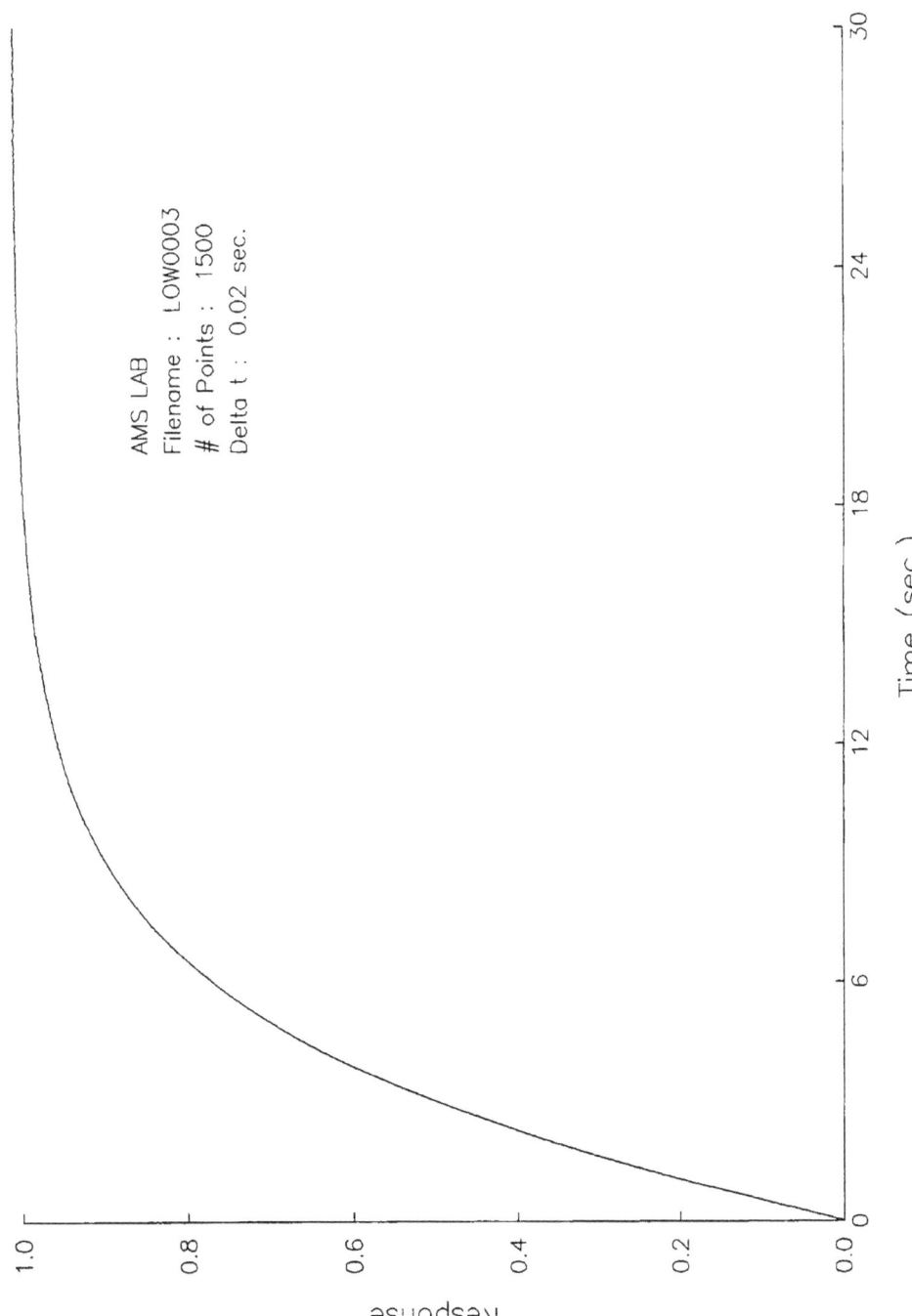

Figure B15. Raw LCSR Transient for Sensor Tag No. N20C .

AMS LAB
Filename : LOW0003
# of Points : 1500
Delta t : 0.02 sec.

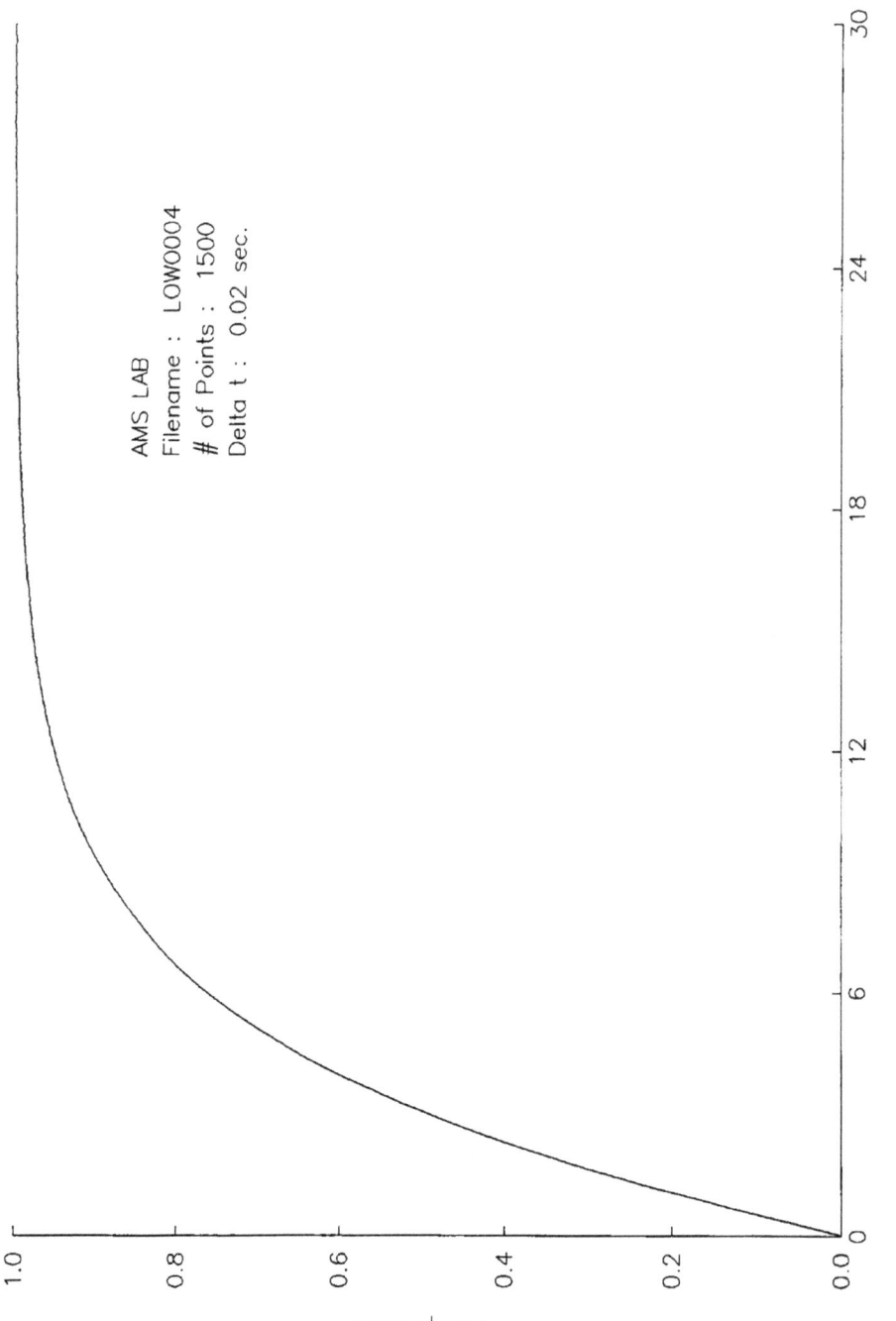

AMS LAB
Filename : LOW0004
# of Points : 1500
Delta t : 0.02 sec.

Time (sec.)

Response

Figure B16.   Raw LCSR Transient for Sensor Tag No. N20D .

Page B16

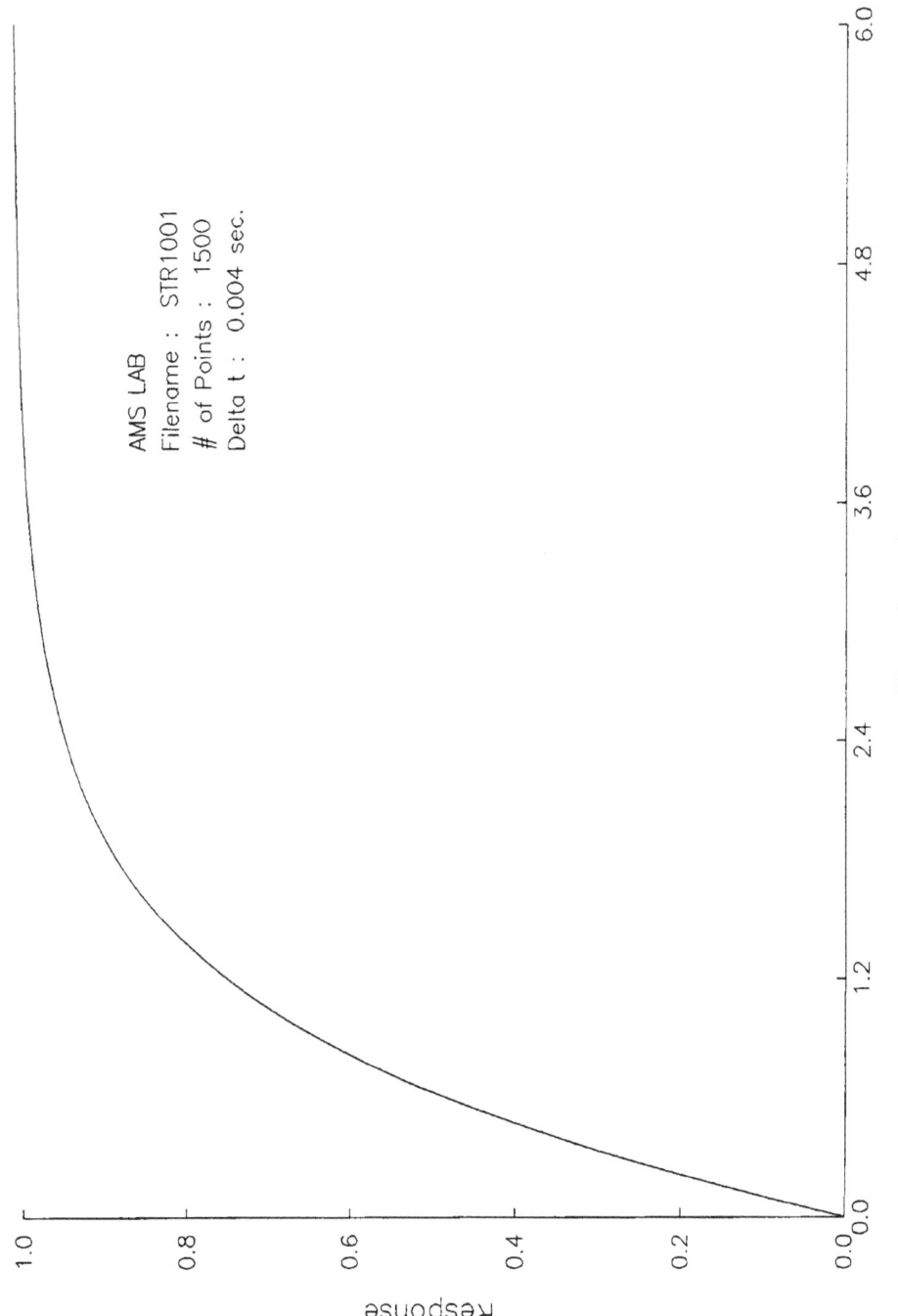

AMS LAB
Filename : STR1001
# of Points : 1500
Delta t : 0.004 sec.

Time (sec.)

Response

Figure B17. A Typical Plunge Test Transient for Sensor Tag No. N5D .

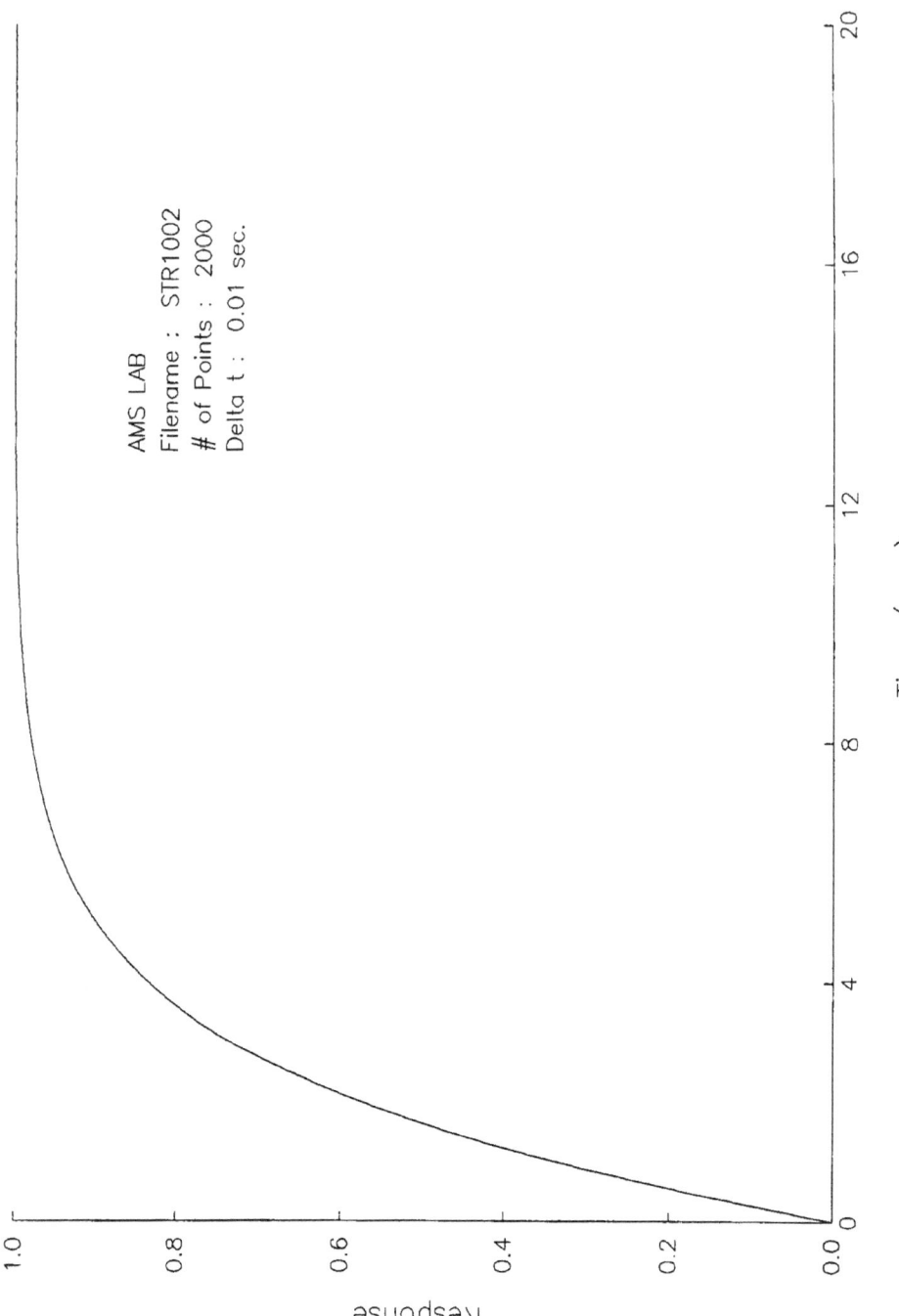

Figure B18. A Typical Plunge Test Transient for Sensor Tag No. N10D .

Page B18

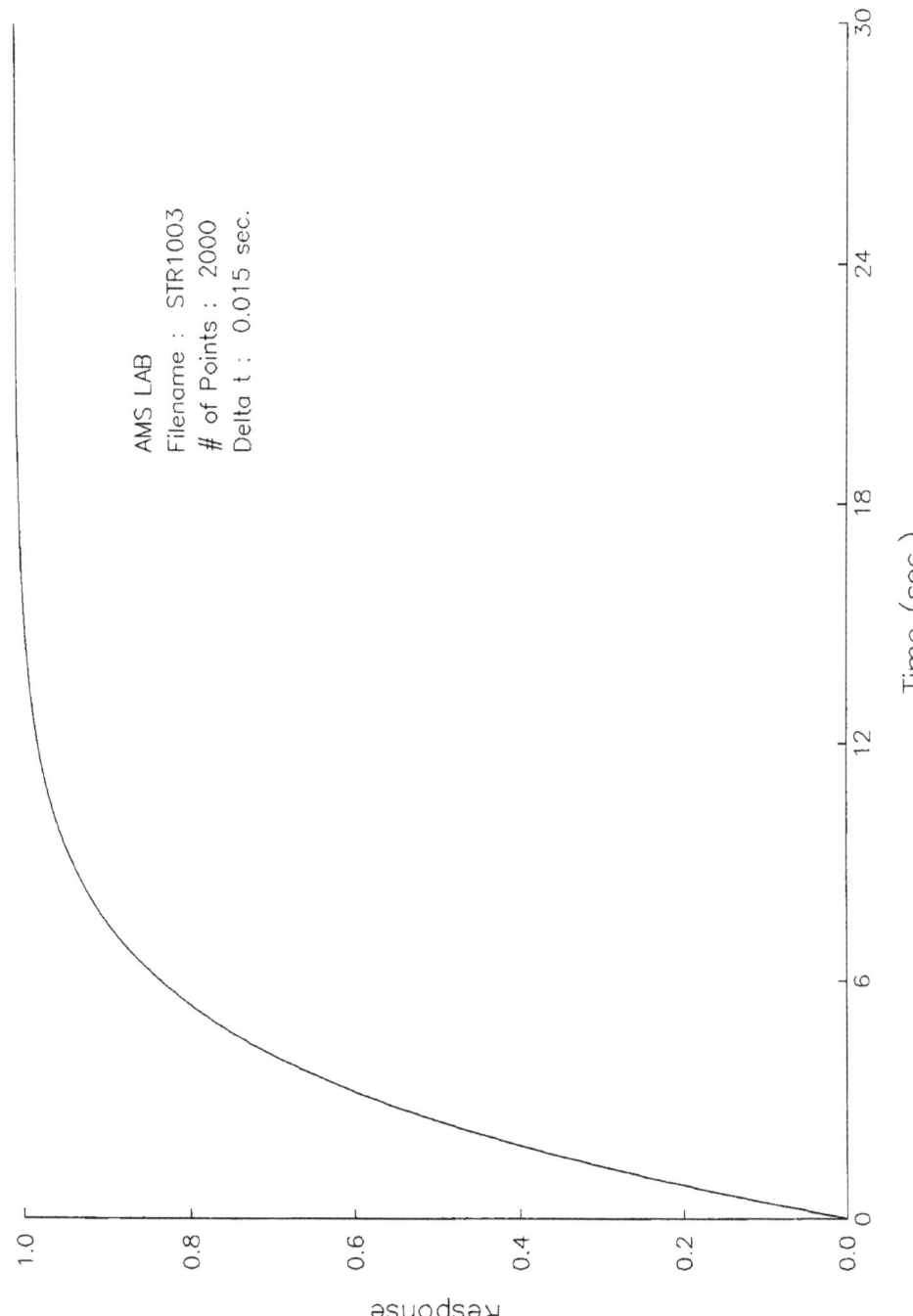

AMS LAB
Filename : STR1003
# of Points : 2000
Delta t : 0.015 sec.

Time (sec.)

Response

Figure B19.   A Typical Plunge Test Transient for Sensor Tag No. N15D .

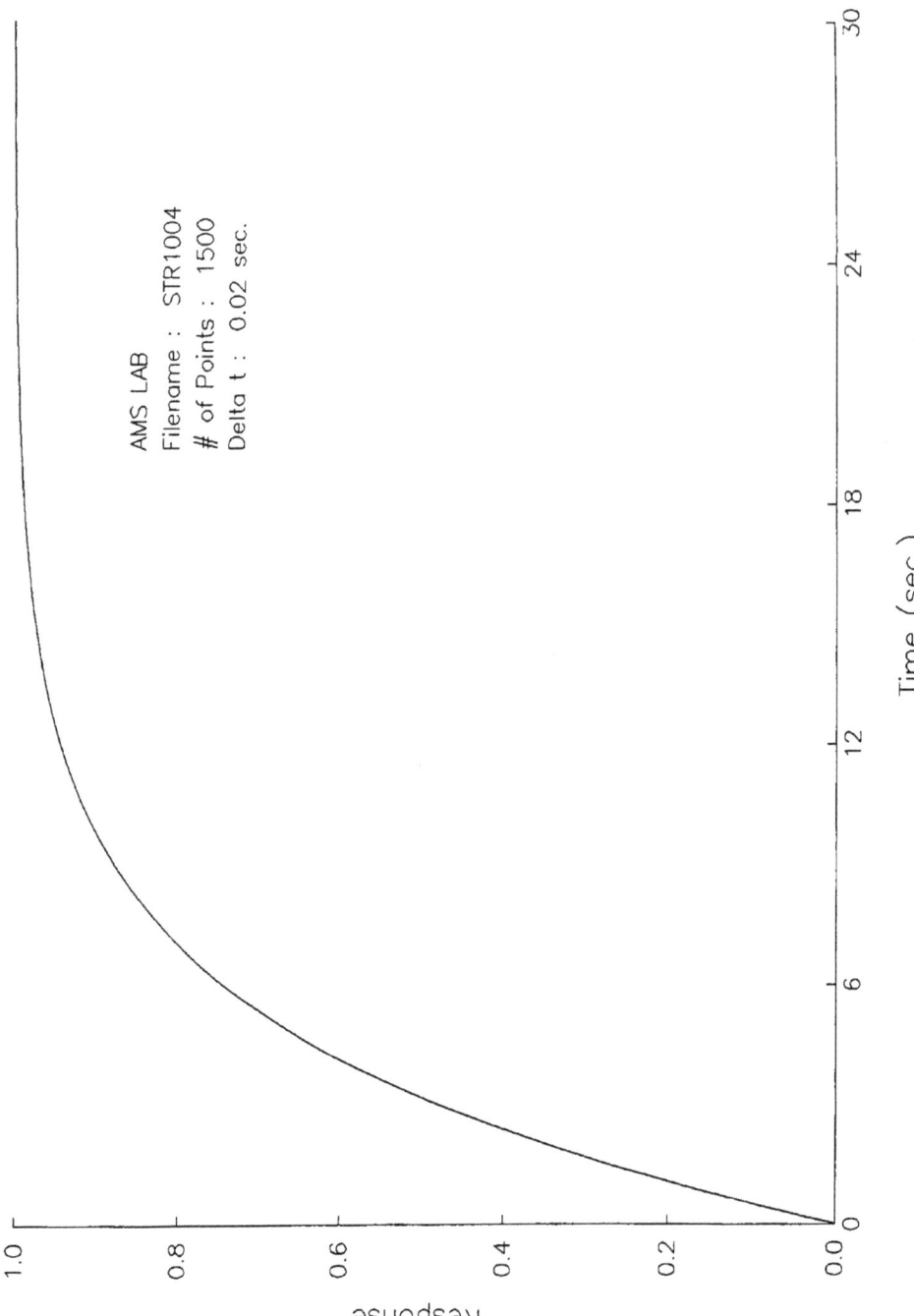

Figure B20.   A Typical Plunge Test Transient for Sensor Tag No. N20D .

NRC FORM 335
(2-89)
NRCM 1102,
3201, 3202

U.S. NUCLEAR REGULATORY COMMISSION

BIBLIOGRAPHIC DATA SHEET

(See instructions on the reverse)

1. REPORT NUMBER
(Assigned by NRC. Add Vol., Supp., Rev., and Addendum Numbers, if any.)

NUREG/CR-6334

2. TITLE AND SUBTITLE

New Sensor for Measurement of Low Air Flow Velocity

Phase I Final Report

3. DATE REPORT PUBLISHED

MONTH	YEAR
August	1995

4. FIN OR GRANT NUMBER

W6378

5. AUTHOR(S)

H. M. Hashemian, M. Hashemian, E. T. Riggsbee

6. TYPE OF REPORT

Technical

7. PERIOD COVERED *(Inclusive Dates)*

10/1/94 - 3/31/95

8. PERFORMING ORGANIZATION – NAME AND ADDRESS *(If NRC, provide Division, Office or Region, U.S. Nuclear Regulatory Commission, and mailing address; if contractor, provide name and mailing address.)*

Analysis and Measurement Services Corporation
AMS 9111 Cross Park Drive
Knoxville, TN 37923

9. SPONSORING ORGANIZATION – NAME AND ADDRESS *(If NRC, type "Same as above"; if contractor, provide NRC Division, Office or Region, U.S. Nuclear Regulatory Commission, and mailing address.)*

Division of Regulatory Applications
Office of Nuclear Regulatory Research
U.S. Nuclear Regulatory Commission
Washington, DC 20555-0001

10. SUPPLEMENTARY NOTES

Copyrighted by Analysis and Measurement Services Corporation, 1995

11. ABSTRACT *(200 words or less)*

This is the report of a six-month feasibility study of a new sensor to measure ambient air flow velocity and direction for health physics applications in nuclear facilities. The information from this sensor is to be used to determine where to place air samplers to sample airborne radioactive material that is representative of the air inhaled by radiation workers. A new sensor was developed in this project and successfully tested in the AMS laboratory for measurement of low flow rates of air. The sensor uses a conventional thermocouple as its sensing element and is therefore referred to as a "thermocouple flow sensor". The dynamic response of the thermocouple is measured using an in-situ response time testing method. The response time information is then converted to a flow signal using predetermined response time-versus-flow correlation for the thermocouple. The thermocouple flow sensor has the potential to aid in determining in-door air flow patterns. This may be accomplished by using multiple thermocouples to measure air flow velocities in several locations in the room and use the velocity information with computational fluid dynamics or neural network models to establish air flow patterns.

12. KEY WORDS/DESCRIPTORS *(List words or phrases that will assist researchers in locating the report.)*

Low Flow Measurement
New Sensor
Response Time Testing
Nuclear Facilities
Radiation Protection
Thermocouples
Health Physics

13. AVAILABILITY STATEMENT

Unlimited

14. SECURITY CLASSIFICATION

(This Page)

Unclassified

(This Report)

Unclassified

15. NUMBER OF PAGES

16. PRICE

Federal Recycling Program

www.ingramcontent.com/pod-product-compliance
Lightning Source LLC
Chambersburg PA
CBHW080251180526
45167CB00006B/2489

* 9 7 8 1 4 9 9 5 7 6 9 5 5 *